Diamond Turn Machining

Theory and Practice

MICRO AND NANO MANUFACTURING SERIES

Series Editor
Dr. V. K. Jain
Professor, Dept. of Mechanical Engineering
Indian Institute of Technology, Kanpur, India

Published Titles:

Diamond Turn Machining: Theory and Practice, by R. Balasubramaniam, RamaGopal V. Sarepaka, Sathyan Subbiah

Nanofinishing Science and Technology: Basic and Advanced Finishing and Polishing Processes, by Vijay Kumar Jain

Diamond Turn Machining

Theory and Practice

R. Balasubramaniam
RamaGopal V. Sarepaka
Sathyan Subbiah

CRC Press
Taylor & Francis Group
Boca Raton London New York

CRC Press is an imprint of the
Taylor & Francis Group, an **informa** business

CRC Press
Taylor & Francis Group
6000 Broken Sound Parkway NW, Suite 300
Boca Raton, FL 33487-2742

© 2018 by Taylor & Francis Group, LLC
CRC Press is an imprint of Taylor & Francis Group, an Informa business

No claim to original U.S. Government works

Printed on acid-free paper

International Standard Book Number-13: 978-1-138-74832-3 (Paperback)
International Standard Book Number-13: 978-1-4987-8758-1 (Hardback)

Visit the Taylor & Francis Web site at
http://www.taylorandfrancis.com

and the CRC Press Web site at
http://www.crcpress.com

Contents

Foreword ix

Preface..xi

Authors xv

1. Introduction...1
   1.1 The Need: Fabricating Smooth Surfaces....................................1
   1.2 Conventional Machining and the Need to Go Beyond 4
   1.3 Diamond Turn Machining (DTM)...5
   1.4 Place of DTM in the Process Chain 8
   1.5 Summary..10

2. Diamond Turn Machines..11
   2.1 Introduction 11
   2.2 Classification of Diamond Turn Machines...............................11
   2.3 Requirements of Diamond Turn Machines...............................12
        2.3.1 Positional Accuracy and Repeatability of Moving
               Elements 13
        2.3.2 Balanced Loop Stiffness...15
        2.3.3 Thermal Effects...16
        2.3.4 Vibration Effects...16
   2.4 Characteristics and Capabilities of Diamond Turn Machines 17
   2.5 Components of Diamond Turn Machines...................................18
   2.6 Technologies Involved in Diamond Turn Machine Building.......20
   2.7 Environmental Requirements for Diamond Turn Machines.......21
   2.8 Sample Machine Specification Sheet...22
   2.9 Summary..22
   2.10 Sample Solved Problems...22
   2.11 Sample Unsolved Problems 25

3. Mechanism of Material Removal...27
   3.1 Introduction 27
   3.2 Comparison of Deterministic and Random Machining Process...28
   3.3 Cutting Mechanisms for Engineering Materials.......................30
   3.4 Micro- and Nano-Regime Cutting Mechanisms.......................35
   3.5 Ductile Regime Machining of Brittle Materials 39
   3.6 Machining of Polymers 40
   3.7 Summary..42
   3.8 Sample Solved Problems...42
   3.9 Sample Unsolved Problems 45

4. Tooling for Diamond Turn Machining... 47
    4.1   Introduction 47
    4.2   Tool Materials and Their Requirements 47
    4.3   Single Crystal Diamond Tools.. 49
    4.4   Tool Geometry 53
    4.5   Diamond Tool Fabrication 55
    4.6   Tool Wear.. 57
    4.7   Tool Setting in DTM 60
    4.8   Summary.. 60
    4.9   Unsolved Problems 61

5. DTM Process Parametres and Optimisation 63
    5.1   Introduction 63
    5.2   Diamond Turn Machining Process and Parametres 63
            5.2.1   Spindle Speed 65
            5.2.2   Feed Rate.. 67
            5.2.3   Depth of Cut 69
            5.2.4   Tool Shank Overhang.. 69
            5.2.5   Coolant 70
            5.2.6   Clamping Method and Footprint Error...................... 71
    5.3   Vibration Related Issues... 72
    5.4   Thermal Issues in Diamond Turn Machining 73
    5.5   Optimization of DTM Parametres.. 74
    5.6   Summary.. 75
    5.7   Sample Solved Problems 75
    5.8   Questions and Problems 77

6. Tool Path Strategies in Surface Generation 79
    6.1   Introduction 79
    6.2   Tool Paths for Symmetric Macro Shapes 80
    6.3   Tool Paths for Producing Asymmetric Macro Shapes.............. 83
            6.3.1   Synchronization of Spindle Rotation 83
            6.3.2   Slow Tool Servo (STS) 84
    6.4   Tool Paths for Producing Micro-Features.................................. 87
            6.4.1   Fast Tool Servo (FTS) 88
    6.5   Tool Normal Motion Path 89
    6.6   Deterministic Surface Generation 90
    6.7   Summary.. 92
    6.8   Questions and Problems 93

7. Application of DTM Products 95
    7.1   Introduction 95
    7.2   Diamond Turn Machining Applications 95

7.3   Applications in the Optical Domain 96
7.4   Polymer Optics Products 99
7.5   Mold Inserts for Polymer Optics...99
7.6   Metal Optics..100
7.7   IR Optics..100
7.8   Diamond Turn Machined Ultra-Precision Components.............101
7.9   Major Diamond Turn Machining Application Areas 101
7.10   Materials Machinable by DTM 102
        7.10.1 Metals 102
        7.10.2 Polymers..102
        7.10.3 Crystals..103
7.11   Summary..103

8. DTM Surfaces – Metrology – Characterization...................................105
8.1   Introduction 105
8.2   Surface Quality...108
        8.2.1   Form Error 108
        8.2.2   Figure Error 109
        8.2.3   Finish Error 109
8.3   Quantification of Surface Errors 109
8.4   Surface Texture 110
8.5   Surface Texture Parametres 112
8.6   Spatial Parametres 115
8.7   Amplitude Parametres 115
8.8   Power Spectral Density 119
8.9   Tolerance..120
8.10   Metrology by Stylus-Based Profilometres....................................121
8.11   Sources of Errors in Surface Quality 122
8.12   Ogive Error..123
8.13   Metrology Errors...124
8.14   Thermal Effects and Metrology 127
8.15   Error Compensation Techniques 128
8.16   Summary..129

9. Advances in DTM Technology 131
9.1   Introduction 131
9.2   DTM Process Monitoring..131
9.3   Developments Related to Machine Tools 133
9.4   Developments Related to Cutting Tools 135
9.5   Influence of Coolant in DTM..137
9.6   Vibration-Based Controlled-Tool Motion.....................................138
9.7   Tool-Path Planning 140
9.8   New Materials and Materials Treatment.......................................142

9.9   Tool Holding for DTM... 144
9.10   Summary... 145
9.11   Questions.. 145

Bibliography.. 147

Index.. 155

Foreword

Any country wishing to get into high-tech manufacturing must develop core strength in advanced manufacturing science and technology. Achieving high precision, in terms of surface, profile and dimensional accuracy, becomes essential for products that depend on high precision and quietness particularly at high speeds, high level of optical performance, molecular level phenomenon and so on for their performance. Sub-micron or even nanoscale precision often becomes necessary in such cases. Diamond turn machining is one of the common and most advanced processes for manufacturing to achieve such high precision.

Diamond turn machining and its deployment for mass manufacture were pioneered by Bhabha Atomic Research Centre (BARC) for its own programmes. Similar developments have also been pursued by other agencies. Today a significant number of diamond turning machines are functional at different institutions and industrial units in the country. This will grow further as the country moves forward with high-tech manufacturing particularly in the context of the 'Made in India' programme. With growing markets in emerging economies, there will be a large demand for low-cost, high-tech products particularly in the form of handheld devices which can provide high-tech services even when highly qualified professionals may not be available. Such products would need a variety of sensors to be incorporated in the handheld devices. This would call for access to critical high-tech manufacturing processes including diamond turning.

I am glad that Dr. R. Balasubramaniam of BARC who has spent a good part of his professional efforts in this area along with Dr. RamaGopal V. Sarepaka, Ex-Chief Scientist, CSIR-CSIO, and Prof. Sathyan Subbiah of IIT Madras have brought out this book on *Diamond Turn Machining: Theory and Practice* that deals with this specialised subject in a comprehensive way. The book will become a handy textbook or reference for a large number of youth that one expects to work in this area.

I am sure the book will prove to be very useful to students, teachers, researchers and industry professionals alike.

Anil Kakodkar
President, National Academy of Sciences, India
Chairman, Rajiv Gandhi Science & Technology Commission
Chairman, Technology Information, Forecasting & Assessment Council

Preface

Manufacturing dates back to the period 5000–4000 B.C. and, thus, is older than recorded history. However, the word 'manufacture' derived from Latin *'manu factus'* meaning 'made by hand' entered the English lexicon in 1567. From these sixteenth century accounts, precision manufacturing has come a long way. In 1974, Prof. Norio Taniguchi coined the word 'nanotechnology', while defining 'ultra-precision machining' as 'the process by which the highest possible dimensional accuracy is achieved at a given point in time'. The keyword is *'at a given point in time'*. With this catchphrase of temporal domain, the nature of precision machining has changed forever, propelling advancements in knowledge and skill-sets (both in human and machine domains, despite limitations of the processes), dictated by ever-growing consumer demands.

Starting from normal lathe operations, precision machining has evolved rapidly by the availability of CNC machines, fast machining algorithms, nanometric controller accuracies, single point diamond turning (SPDT) machines and auxiliary systems. These fabrication advances are supplemented by advances in metrology equipment and techniques to deliver work-pieces of hitherto unachievable surface accuracies. In this journey, diamond turn machining (DTM) has emerged as a fully matured precision machining process, increasingly deployed by a growing population of precision component and system developers. Though SPDT has been around for more than 5 decades, its significance has caught users' attention in the last 35 years.

DTM is a near perfect marriage of ultra-precision vibration-free equipment, stiff tool holder and a well-chosen fixture, a diamond tool of prescribed geometry and orientation, to initiate and maintain a calibrated material removal rate with minimal cutting forces, to deliver a precision component with the desired surface accuracies within agreed tolerance ranges.

Despite DTM's wide acceptance and regular practice by the ultra-precision machining community, the lack of a systematic study material is sorely felt by the beginners and practitioners of this specialized craft. This book attempts to meet this requirement by presenting an overall view of various facets of DTM, while providing critical insights from various aspects of DTM operations.

The material in this book is structured into nine chapters, with each as a building block with inputs towards the deliverance of precision components.

Starting with a discussion on the need of precision machining and with an introduction of DTM in Chapter 1, the reader is provided with vital insights (Chapter 2), into the world of diamond turn machines in terms of their classifications, along with their features, limitations, operational conditions and the technologies involved in the building and operation of DTM equipment. Chapter 3 discusses the deterministic approach and material removal mechanisms, while comparing the macro- and micromachining scenarios.

The DTM processes need to be complimented by the suitable diamond tools to deliver ultra-precision surfaces with submicron waviness and nanometric roughness. Chapter 4 presents an in-depth discussion on diamond tools, their geometries and tool-wear.

After an overview of DTM machines and diamond tools, the reader is taken on an itemized tour of DTM process parametres, their influence on work-piece, process parametric optimisation, vibration and thermal issues in Chapter 5. This is followed by methodologies on improvement of surface quality by tool path management of conic and complex surface profiles in Chapter 6.

A detailed discussion on the application areas of DTM including societal, commercial, strategic and biomedical instrumentation is included in Chapter 7.

Precision fabrication and precision metrology are Siamese twins and these processes need to complement each other. Chapter 8 introduces the surface quality criteria in terms of form, figure, finish and related quantification parametres, with a discussion on the genesis of surface errors qualitatively and on the error correction philosophy.

The Chapter 9 is a peek into the future of DTM. The global DTM community is growing with novel challenges coming its way, in terms of complex surface shapes, large size work-pieces, surface qualification practices and cost-effectiveness.

The authors, due to their diversified backgrounds (academia, R&D and industry), have presented the subject matter heterogeneously, with a common goal to introduce DTM to the beginners and to the practitioners alike. The sole purpose of this book is to present systematically various topics of DTM as a primer of DTM. In-depth discussion and relevant case studies of each aspect of DTM beg a larger canvass, maybe at a later date. Therefore, this book does not claim to be the final word in this domain. It intends to serve several audiences while introducing each to the other, with the anticipated appreciation for the craft and science of DTM.

The authors are indebted to various practitioners of the precision instrumentation community, who are too numerous to name individually: mentors and teachers; funding agents and authorities; colleagues and students; vendors and customers; families and friends; and finally supporters and naysayers. The authors want to express their special thanks to their respective family members for their patience and encouragement to enable this book project. As the work presented belongs to all above, they alone deserve all the credit. However, the authors claim sole responsibility for all oversights and errors in presenting this labour of love to the precision component development community.

Knowledge in this niche area will evolve further in coming years and better books will be written for and by a mature DTM community.

The authors eagerly await that event.

R. Balasubramaniam
RamaGopal V. Sarepaka
Sathyan Subbiah

Authors

Dr. R. Balasubramaniam earned his PhD from IIT Bombay (2000). He is working in the field of ultraprecision manufacturing (UPM) for multiple application areas for the last three decades at Bhabha Atomic Research Centre, Mumbai, India. Apart from R&D activities, he is active in teaching, guiding and mentoring doctoral and graduate students in various aspects of UPM and associated technology domains. Additionally, Dr. Balasubmaniam is involved in the promotion of UPM through academia and industry for societal applications including agriculture, assistive devices, health and education for rural areas. His other significant activity involves the growth of the diamond turn machining community both in the country and globally. He has more than 100 publications to his credit apart from contribution chapters on diamond turning, nano-finishing and micro-turning in textbooks on micro-machining.

Dr. RamaGopal V. Sarepaka's career graph includes academics at IIT Delhi; tenures at CSIR-Central Scientific Instruments Organisation (CSIR-CSIO) (Chief Scientist) and Academy of Scientific Innovative Research (CSIR-AcSIR) (Professor); as visiting scientist at the Centre for Applied Optics, University of Alabama in Huntsville, Alabama, and at the College of Optical Sciences, University of Arizona, Tucson, Arizona. After his R&D and academic stints, Dr. Sarepaka is currently working with the single point diamond turning (SPDT) based precision optics industry in India. His research interests include precision optical instrumentation, SPDT and surface characterization, optical fabrication, optical design, tolerance analysis, with a sizeable volume of published work. He has guided several PhD students and graduate students in SPDT and aspheric optics.

Dr. Sathyan Subbiah is an associate professor in the department of mechanical engineering at the Indian Institute of Technology Madras. He earned his PhD from Georgia Tech in 2006, MS from the University of Illinois Urbana Champaign in 2000, and B.Tech from IIT Madras in 1997. He worked for three years in the automotive industry in Detroit where he had his first opportunity to use diamond turning for an automotive component. His more elaborate research ventures into diamond turning came during his time as an assistant professor at Nanyang Technological University Singapore. Here, he and his students attempted some unconventional ways to produce surface textures in diamond turning and also to improve flatness at edges. He is now actively exploring research opportunities in India in this area. He has guided three PhD students and several students in Masters programs in the broad area of machining.

1

Introduction

1.1 The Need: Fabricating Smooth Surfaces

We interact every day with an optical component in the morning – the ubiquitous common mirror. Made of flat float glass with a smooth surface and a metallic reflective coating at the back, it faithfully reflects visible light. Having a smooth surface (obtained using gravity by floating liquid glass in a molten tin bath) is important – so that the interacting optical rays don't scatter and lose their intensity and ability to focus both from the front surface and also the back surface of the glass. How smooth is smooth enough for reflection/transmission? Well, the surface geometrical undulations have to be of an order lower than that of the wavelength of the light (Figure 1.1).

Engineering applications often involve more than just visible light ranging from larger wavelength infrared to lower wavelength X-rays. The lower the wavelength of electromagnetic (EM) waves used (e.g. X-rays with wavelengths less than a few nanometres), the more demanding is the surface smoothness – surface undulations of the order of Angstroms for X-ray mirrors. Engineering applications also involve complex surface shapes (Figure 1.2) requiring EM wave reflection from the interacting top surface without passing through glass, like in the common mirror.

Such engineering applications demanding surface smoothness include:

- Defense forces require specialised instruments that use the infrared (IR) part of the EM spectrum for night-vision. Other defense products include helmet-mounted display systems, head-up systems, virtual reality systems, avionics and other imaging systems used in air and land defense systems.

- X-ray beam deflections also require the use of highly finished curved surfaces in silicon (Figure 1.3). Diamond turning is an important technology that forms a key precursor to the final step of optical polishing involved in making such mirrors. Such mirrors are made today, for example, by companies like Zeiss in Europe.

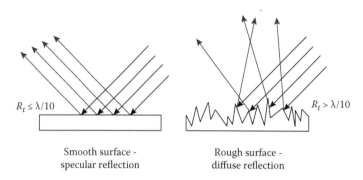

FIGURE 1.1

EM wave reflection requires surface undulations of one order lower than that of the wavelength of the impinging EM wave. Transmission also requires such surfaces on both sides of the component.

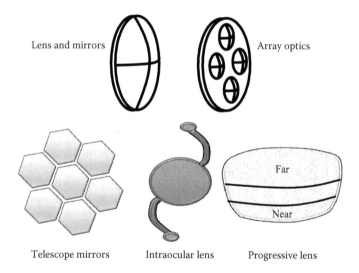

FIGURE 1.2

Complex geometries with demanding surface smoothness needed in a variety of applications from space and defence to medical.

- Diamond turning technology is also needed to meet the requirements of the common human: medical products are needed to aid vision – both outside the body and implants for use inside the body. Such optical elements are typically made of polymers and are either directly processed using diamond turning or indirectly in the form of diamond-turned-inserts used in injection molds for mass replication. Many companies make such products commercially – ophthalmic implants, mobile camera lens and progressive lenses.

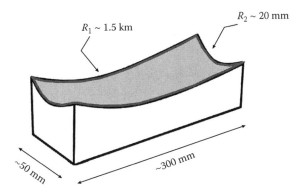

FIGURE 1.3
Complex shape needed for a typical X-ray mirror.

- An important application of diamond turning is in manufacturing large mirrors needed for telescope applications. Such telescopes are often made by a series of international collaborative agreements by several interested countries.

- Space technology also requires diamond turning technology for making several specialised rocket components that require high precision smooth surfaces. Recent avionic requirements are exploring precision surfaces from titanium substrates.

- Medical products such as many structural implants require very smooth surfaces for optical, tribological reasons and also for better integration with the body.

The requirements of reflection, transmission, index of refraction, dispersion and index gradients also demand the use of varied engineering materials (Table 1.1).

The challenge now is how to process such a varied list of materials – some ductile, some brittle, some soft, some hard – to produce optically smooth surfaces that are of complex geometry. The manufacturing process that is, in general, known to produce precise dimensions, tolerances and surface smoothness is machining and abrasive grinding/polishing. The next section

TABLE 1.1

Materials to be Processed for Fabricating Components Interacting with EM Waves

Visible Spectrum	Infrared Spectrum	X-Ray Beams
Metallic alloys (Al, Cu)	Flourides	Silicon
Polymers	ZnSe and ZnS	
	Silicon, Germanium	
	Chalcogenides	

considers the common machining processes and shows how they are insufficient in producing such demanding smooth surfaces and hence modifications are in order to create a new process – the DTM.

1.2 Conventional Machining and the Need to Go Beyond

The reader is probably aware of the common machining processes: turning, milling, drilling, boring etc. These are basic shaping processes that shape a raw material by removing unwanted material and thus the substrate attains the needed shape. In this process of material removal, they create new surfaces on the material. Each of these processes uses cutting tools that move relative to the work material surface, impact it and remove material by a brute force method. The precision with which the tool can be moved in 3D space enables the carving out of complex shapes with good dimensional and tolerance control. The process involves a rigid machine that holds both the tool and the work rigidly, while providing needed controlled cutting and feed motions to one or both of them.

The way the material reacts to the shear/fracture involved during this tool–work interaction, the way the tool moves with associated irregularities of motion to various areas of the surface (called feed motion), the cutting motion action and the macro/micro geometry of the tool, all leave behind marks and surface undulations on the work surface.

Besides the nature of the work material and its properties, the sharpness of the cutting tool is an important factor that dictates the nature of shear-fracture interactions at the tool cutting edge. Typical sharpness of tools (made of hard multigrain carbides, ceramics and polycrystalline diamond and cubic boron nitride) used in conventional machining are of a few to tens of micrometres. Besides the sharpness, the tool edge profile itself is hard to control in these tool materials and leaves a very strong signature on the work surface. In order to provide complex motions to the tool and the work-piece, the machine tool used has various linear/rotary slides that often involve rigid metal-to-metal contacts to provide the needed stiffness of motion. Due to design constraints, often considerable vibrations are inevitable and these vibrations are built up over several passes on a surface. Besides, these slides and joints offer poor damping to suppress the vibrations quickly. All these lead to variations in tool edge motions, further causing surface marks and undulations. Several efforts have been made to reduce the effect of these factors, thus leading to the best surface finishes possible in precision machining [1] of peak-to-valley roughness (R_t), of several micrometres (average R_a) in the range of submicrometres – a far cry from that needed for IR, visual and UV/X-ray optical components. Typical surface requirements needed for these components are ±0.1 μm of machining precision, a surface finish of better than 0.01 μm Ra, and 10^{-6} as the ratio of error tolerance and machined dimension [2,3]. The graduation from conventional precision machining to the next realm of ultra-precision machining (UPM) is temporal [4] and can only be an outcome of smart manipulation of modified tooling, state-of-the-art high-precision machining equipment, processing algorithms, environmental cleanliness and maintenance and implementation of error compensation procedures to deliver components meeting the stringent quality criteria.

Hence, there is a need to go beyond conventional machining processes and look for drastic changes holistically, in all aspects of machining so that the requirement of surface smoothness can be addressed effectively. This leads us to the new process of diamond turn machining (DTM).

1.3 Diamond Turn Machining (DTM)

Diamond turn machining (DTM) is a class of ultra-precision machining processes that holds a unique place within the domain of single-point mechanical material removal processes such as turning, milling, drilling etc. The uniqueness comes from the condition of the machined surface produced by diamond turning: surface roughness typically of the order of a few nanometres, surface form tolerance of the order of fractions of micro-metres and now the increasing ability of this process to produce micro-structured (textured) surfaces such as in diffractive optical elements. Such surface conditions have been achieved using diamond turning in a broad class of materials ranging from ductile to semi-ductile to even very brittle materials such as silicon.

The near ideal marriage of the multiple matrix interaction of the precision diamond tool, facilitating controlled lathe operation with precision material removal, exacting protocols of machining and equally demanding regime of surface characterization, coupled with a deterministic approach will result in the desired surface quality of nanometric surface roughness and submicron surface waviness to meet near theoretical standards in a system whose performance is severely limited by budgets on its weight, volume and footprint for neo-compact precision modules of next-generation precision instrumentation.

The DTM surface quality is routinely characterized in terms of fraction of submicrons (for its surface profile departures from prescribed datum lines) and nanometres (for its surface roughness) [5]. Achieving such fine surfaces requires a combination of some unique characteristics of the otherwise typical machining process elements comprising the cutting tool, process conditions, machine tool, motion control, fixture and measurement. It includes;

- To affect a smooth shearing action, a very sharp and uniform cutting tool edge, with waviness and edge radius of the order of a few to tens of nanometres, is essential. Such fine edge radius and waviness are possible to achieve only in hard single crystal materials such as in diamond, wherein the desired crystallographic planes and directions can be chosen with care to carve the edge – a process that requires considerable expertise for indigenous development and also which increases tooling costs.

- Besides the fine cutting edge, it is also essential to maintain smooth chip formation which will result in a machined surface devoid of indentation effects and fracture, the latter being prominent in brittle materials, wherein ductile-regime machining conditions are essential. Ductile-regime can be maintained by keeping all dimensions of the chip load cross-section below a critical level and also by inducing compressive stresses via the cutting edge radius and negative rake angles.

- To affect the required cutting motion and feed motion, the tool and work materials have to be mounted with minimal holding stresses, on vibration-free/damped high precision sliding bearing systems, both linear and rotational (e.g. spindle). One of the best ways to achieve such smooth motion is to remove mechanical contact between the moving elements by separating them with a small gap filled with temperature-controlled pressurised fluids such as air or oil. Such specialised bearing systems pose a challenge to indigenously design and manufacture and hence lead to high capital investment.

- Another unique characteristic of diamond turning that has emerged recently is the simultaneously coordinated motion, at relatively high speeds, of the spindle motion and z-direction feed axis. This leads to a chiseling-like dynamic motion along the tangential cutting path resulting in the creation of interesting textured features such as that needed for diffraction gratings. Such coordinated motion, while typically avoiding milling-type intermittent motion, causes undesirable overlapping of the returning tool path on the already machined surface. The smart use of such coordinated motion requires complex controls that are built-in and programming strategies that typically need to be custom developed for a particular texture to be made.

- Diamond turning also requires specialised high-precision fixtures to be developed for holding the work material (e.g. using a vacuum) and cutting tool (for fine adjustments). The dynamics of the fixtures also play an important role in reducing the vibration effects associated with the process. Any error either due to the diamond tool holding or in the tool dynamics will lead to damage and gross deterioration of the surface quality of the profile of the work-piece under process and will shorten the life of the precious precision diamond tool as well.

- The fine surface generated requires high precision metrology systems that can verify whether the needed tolerances have been attained and possibly, close the loop, in-line or off-line, with the diamond turning for process condition adjustments, should there be a gap (which is most common in all conditions) between the desired and achieved surface geometries.

The first specialised machines for diamond turning were built in the mid- to late 1970s in the United States by Lawrence Livermore National Labs. They have built both small and very large DT machines (DTMs). Since then, the technology has come a long way with commercial products now readily available for purchase largely from the Untied States, United Kingdom and Japan. Research in the DTM area is continuing worldwide with reports coming in from Japan, Hong Kong, India, China, South Africa, United States, United Kingdom and other countries in Europe.

As is the case with all revolutions in instrumentation, the philosophy of diamond turning is relentlessly driven by the necessity of applications in terms of shrinking design rules of electronic circuitry, progressively contracting geometries of the precision components, ever-increasing global demands on miniaturisation of commercial products and a mandate for large-volume productions (to maintain the healthy bottom lines of companies). These commercial aspects of DTM development are providentially matched by major advances in controls, feedback systems, servo drives, and general machine design and construction (in terms of stiffer axes, smoother drives, and more precise spindles), tooling (by smart manipulation of diamond crystal geometry) and deterministic micro-machining approaches. This machining innovation has led to the explosion of previously unknown surface shapes and profiles in terms of plan and spherical surfaces leading to conics, diffractive elements, torics and freeforms, with the surface figure errors well within the submicron ranges and the surface roughness reduced up to single digit nanometres. These innovations are the product of evolution of a host of new technologies. The resultant module accuracies and their surface quality have been the result of a regimented machine approach, often termed the deterministic approach. This approach directly addresses the randomness of the operation and brings more certainty by clear monitoring and control of the variables involved [6]. The global phenomenon of merger of different knowledge domains for advanced instrumentation has left its mark on DTM processing as well. Researchers from very dissimilar backgrounds of science and technology have joined hands through DTM [7]. The application areas have necessitated this happy union of skills across many domains. One of the happy marriages of knowledge, skill and outcomes is between ultra-high-precision machining using DTM and optical instrumentation development [8]. In fact, the DTM-based aspheric shape development has erased the multiple restraints of conventional optical fabrication technologies and has provided the research community, industry and thereby society with many options of system development in a multitude of application areas [9,10].

It must be mentioned here that, due to the involvement of state-of-the-art equipment, novel diamond tools, uncompromising machining process, stringent qualification criteria and expensive metrology, development of ultra-precision components is a very expensive proposition. Despite this major limitation, the optical production industries have adopted DTM processes by fine-tuning their process flow-charts and by optimal utilisation of resources, for assembly line productions with spectacular outcomes. In last two decades, the global DTM activities have increased significantly, with a multitude of applications deploying the DTM route to meet their respective objectives.

The dynamics of an effective DTM operation broadly include: material composition and properties of the work-piece, status and operational environment of the DTM equipment, condition of the equipment, dynamics of diamond tool geometry (nominal and actual), machining parametres, machining conditions, complexity of the work-piece to be processed, type of metrology equipment deployed and its capabilities/limitations, machining and metrology protocols established and maintained, environmental conditions during machining and metrology stages, fabrication and metrology skills available, appropriate analysis of the results obtained, available tolerances for the prescribed specifications and the application where the precision component is being used. Apart from these aspects, other matters like setting of the diamond tool, DTM equipment dynamics, thermal issues during DTM, etc., [11] also affect the outcome of the effort ominously. It is thus explained in the subsequent chapters in detail. However, it may be mentioned that one may aim at obtaining a complete picture of DTM operations and associated issues, only by an active pursuit of the development of precision components by DTM.

To summarise, diamond turn machining is a niche machining process holding a unique forte in the field of machining. This section highlighted various aspects of this uniqueness. It is also noted here that the name – diamond turning machining (DTM) – is rather unique and introduced here in this book. The alternate, perhaps more popularly used, names the reader will find in the literature are 'diamond turning', 'single point diamond turning (SPDT)' and 'ultra-precision machining'. These processes are the same as that referred to here as DTM. This book will retain this new name of DTM throughout all the chapters.

1.4 Place of DTM in the Process Chain

Components that need a high level of surface smoothness typically incorporate the DTM process. The process chain to make this product starts from making the rough shape using conventional bulk processing techniques (e.g. casting).

A further semi-rough shaping is carried out using conventional machining processes of milling and/or turning. The DTM process then provides the needed form and surface (Figure 1.4). A subsequent polishing process may be required depending on the type of EM wave with which the surface is expected to interact.

As mentioned earlier, the requirement of surface smoothness is related to the nature of the EM wave that is interacting with the surface. Component surfaces that deal with infrared (IR) waves can be processed by DTM alone (Figure 1.5). The mid-spatial frequencies seen on a DTM surface do not pose any major problem for these applications. Components that require visible light reflection and transmission cannot tolerate the mid-spatial frequencies

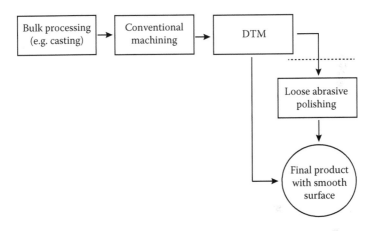

FIGURE 1.4
Typical process chain for manufacturing optical components.

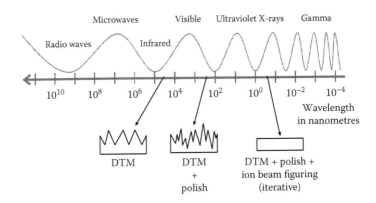

FIGURE 1.5
Surfaces dealing with infrared waves can be processed via DTM, while those dealing with visible light require subsequent polishing and surface deflect X-ray beam require several iterative rounds of polishing and measurement after DTM processing.

arising out of the deterministic cutting and feed motion paths seen in DTMs; these components require subsequent loose abrasive polishing processes that introduce isotropism in the surface. The DTM process provides the much needed form control of the surface, while the polishing processes provide the local surface smoothness. Optical surfaces for X-ray instrumentation demand high levels of slope and roughness control. These surfaces undergo several rounds of iterative polishing and measurement cycles subsequent to DTM, sometimes requiring special processes such as ion-beam figuring for fine material removal. Again, the DTM process helps achieve the overall geometrical form.

1.5 Summary

This chapter introduced the reader to the diamond turn machining process. Starting with the main driver of producing smooth surfaces on various materials (metallic, polymer, semiconductors) this chapter described how the historically prevalent conventional machining methods were insufficient and hence were modified – from machine structure, to cutting tool material and geometry, and motion paths and associated control – leading to the development of ultra-precision diamond turn machining (DTM) as a niche process to produce optically smooth surfaces – both flat and complex ones. The DTM process – also called the single-point diamond turning or ultra-precision machining – has certain distinct advantages of forming overall shapes over polishing to make components. It is thus explained to the reader why DTM has the place it occupies in the process chain now widely adopted to make demanding optically smooth surfaces.

2

Diamond Turn Machines

2.1 Introduction

Diamond turn machines (DTMs) are ultra-precision machines having the capability to generate surfaces with nanometric levels of accuracy and precision. These surface quality parametres are affected by the machine quality, which is controllable at the machine building stage, as well as by the inaccuracies arising out of process variables. Hence, building machines with very high levels of accuracy becomes the first step in the journey towards achieving ultra-precision machined surfaces. Even though machine building activity is well understood and established worldwide, there are only a few diamond turn machine builders. Stringent demands on accuracy levels have confined the knowledge for building diamond turn machines to a very few manufacturers.

An attempt is made in this chapter to provide insights relevant to diamond turn machine building. Initially, various types of diamond turn machines are presented. Subsequently, characteristics and capabilities of the diamond turn machine, its components, technologies involved and the environmental aspects affecting the performance of diamond turn machine tools are discussed. To have an overall idea, a typical specification sheet of a diamond turn machine is also presented.

2.2 Classification of Diamond Turn Machines

Like most of the machines, diamond turn machines are also classified based on their number of axes and their configurations. Following is one of the popular ways of classifying diamond turn machines.

Type A: X, Z Lathe machines

Type B: X, Z, C Lathe machines

Type C: X, Z, C, B Lathe machines

Type D: X, Y, Z, A, B Milling machines

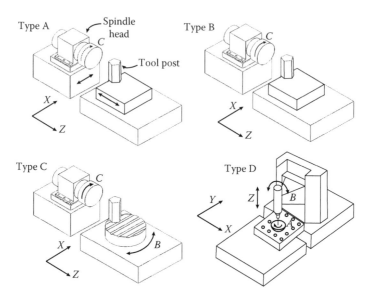

FIGURE 2.1
Classification of diamond turn machines.

Figure 2.1 shows the schematics of the same.

Type A machines are similar to conventional lathe machines wherein, both the X-axis, which carries head stock with spindle, and the Z-axis carrying the tool can be programmed for simultaneous movements and axisymmetric features can be generated. With appropriate fixtures, off-axis parabolic surfaces can also be generated on this machine.

Type B machines have additional control on the spindle (C-axis) and this enables it to generate non-axi-symmetric features. When used along with attachments like fast tool servo, multiple features like lenslet arrays can be generated on this type of machine.

Type C machines are provided with control on the B-axis, which holds the cutting tool and enables it to remain normal to the surface being machined. For spherical surface machining, this type of machine is most preferred.

Type D machines are similar to milling machines; however, these machines employ a fly tool mounted on a spindle.

2.3 Requirements of Diamond Turn Machines

The ultimate requirement of an ultra-precision machining process is to generate the desired surface profile with deviation in the order of a few nanometres (nm) or less. The lesser the deviation on the actual profile of the

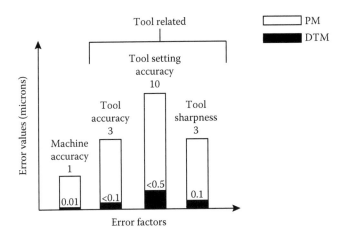

FIGURE 2.2
Factors contributing to the inaccuracies on generated surfaces.

surface from the targeted profile, the better is the process. In case of diamond turn machining, deviations in the order of a few tens of nm are targeted and to achieve this, all the aspects including machine building are given special attention. Figure 2.2 shows the major elements contributing to the deviations of the machined surface both on conventional precision machines (PM) as well as on diamond turn machines.

When conventional precision machines can generate surfaces with accuracy level better than 1 μm, diamond turn machines can generate a similar surface within a few tens of nm accuracy. The contribution of a machine tool's inaccuracy is much less in the case of diamond turn machines compared to conventional precision machines [12]. Inaccuracies due to tool and process variables are well controlled and compensated to a larger extent in diamond turn machines; however, to have extreme accuracy in diamond turn machines, special attention is required during machine building.

Accuracy of diamond turn machines depends on the following factors in a significant way:

- Positional accuracy and repeatability of moving elements
- Balanced loop stiffness
- Thermal effects
- Vibration effects

2.3.1 Positional Accuracy and Repeatability of Moving Elements

Positional accuracy is the degree of agreement between the targeted value and the programmed value of the moving slides and spindle; repeatability is its ability to reach the same position in a repetitive manner. When the degree of agreement in the order of a few tens of nm is achieved, one of the major objectives of diamond turn machine building is met. In general, positional accuracy and repeatability of the diamond turn machine is 2 to 3 orders better than any precision machine. Positional accuracy and repeatability of diamond turn machines are affected by the following factors:

- Degree of freedom of moving elements
- Geometrical accuracy of the axis of moving element and its datum
- Friction between moving elements
- Scale, drive and feedback elements

Degree of freedom of any moving element only in the desirable direction ensures elimination of errors. For example, X-axis movement through the guide-ways should ensure freedom of movement in the X-direction (translation) only and all other degrees of freedom should be restrained. Figure 2.3 schematically shows the various degrees of freedom in all 3-axes and desirable degrees of freedom for the X-axis and spindle, respectively. In Figure 2.3a, the major sources of errors arising due to unconstrained degrees of freedom other than translation in X-axis are the clearances between the moving components, geometrical errors in the sliding surfaces of the X-axis and friction between the moving parts. Hence, the design should ensure that the 5-degrees of freedom, namely: Y, Z, ω_x, ω_y and ω_z should be eliminated to achieve accurate X-axis movement; in a similar way for other axes also, the undesirable degrees of freedoms should be constrained. Various errors

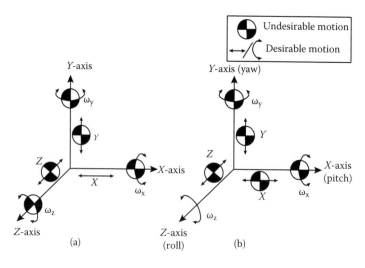

FIGURE 2.3
Schematic diagram showing desirable degrees of freedom for (a) X-axis slide and (b) spindle.

attributing to the machine spindle inaccuracies including pitch, yaw and roll are shown in Figure 2.3b. Errors related to datum inaccuracy, scale and feedback system also affect the positional accuracy of the machine.

2.3.2 Balanced Loop Stiffness

Loop stiffness in a diamond turn machine indicates the equivalent stiffness values of different machine elements during machining. Elements forming the loop stiffness in Type A machines can be represented as follows:

> – Work piece – fixture – spindle – headstock – X-axis table – X-axis guide-ways – bed – Z-axis guide ways – Z-axis table – tool post – cutting tool – work piece.

The cutting force F acting at the tool – work piece interface is related to mass (m), acceleration ($\ddot{x}$), damping coefficient (c), velocity ($\dot{x}$), stiffness (k) and deflection (x); and it is represented by the following equation:

$$F = m\ddot{x} + c\dot{x} + kx \tag{2.1}$$

Joint stiffness between any two elements can be represented by springs with corresponding stiffness values and the bearing elements by damper as shown in Figure 2.4. The force generated during the machining process at the tool–work piece interface is transmitted through these elements in both directions. Various elements deflect to different amplitudes for the same force and cause an unbalanced loop stiffness resulting in changed tool path of the cutting tool.

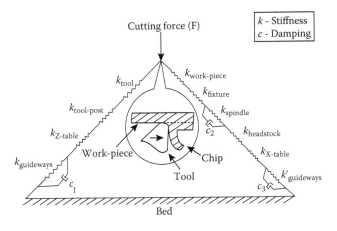

FIGURE 2.4
Loop stiffness of various elements of a diamond turn machine.

As the loop stiffness of the machine is affected by the individual stiffness and damping values of different machine elements, any variations in them result in generating vibration and chattering on the machined surface. For example, when the 'clamping length' of the tool shank is changed from optimum value, it changes the 'tool overhang value' and subsequently the stiffness value of the tool shank. The resulting deflection at the tool tip causes chattering on the machined surface. Hence, selection of material and their geometries for different machine elements to form the balanced loop stiffness becomes an important design consideration while building diamond turn machines. Since the slides and bearings also provide damping, their characteristics play a major role in the loop. For example, damping characteristics of different bearings like sliding surfaces, linear motion guide-ways, aerostatic bearings and hydrostatic bearings vastly differ from each other and selection of any one type of the above-mentioned bearings affects the selection of other elements in the loop stiffness.

2.3.3 Thermal Effects

Thermal drift significantly affects the performance and accuracy of the diamond turn machines. Due to the heat generated by drivers, friction generated by the rotation of the spindle and movement of the slides and cutting process, thermal drift takes place and it causes differential expansion of various machine elements. Expansion due to thermal drift causes many undesirable effects, including altering of the clearance between bearing surfaces of the spindle as well as table slides, axial growth of the spindle, headstock height growth, tool shank length change, etc. Spindle growth in the axial and radial directions affects both the position and orientation of the rotational axis and subsequently the size and shape of the component. Proper selection of material and cooling arrangement enables minimisation of thermal drift. In the present-day diamond turn machines, temperature sensors are embedded at various locations and compensation strategies are used to minimise the effect of thermal drift.

2.3.4 Vibration Effects

One of the major requirements of diamond turn machines is to generate optical quality surfaces which are free from size and shape deviations as well as from cosmetic defects like scratches, digs, chatter marks, etc. Vibration at the interface of tool and work-piece is one of the major reasons for generating such defects on the machined surface. In addition, vibration affects tool life significantly.

The resulting vibration at the interface of tool and work-piece is due to

- Tool and tool tip vibration
- Spindle vibration
- Material-induced vibration
- External vibration

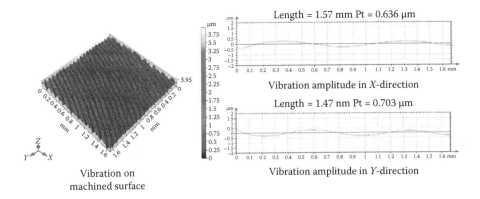

Length = 1.57 mm Pt = 0.636 μm

Vibration amplitude in X-direction

Length = 1.47 nm Pt = 0.703 μm

Vibration amplitude in Y-direction

Vibration on
machined surface

FIGURE 2.5
Effect of vibration on the machined surface.

Vibration causes relative displacement in a periodic manner between the tool and work-piece. This undesirable displacement causes deviation in the desired tool path at a microscopic level and hence the topography of the generated surface and its surface finish are affected. The tool shank is a cantilever, so the free vibration at its tip causes uncontrolled relative displacement. Similarly, spindles of diamond turn machines are supported either by aerostatic or hydrostatic bearings. They are subjected to shift in axial and radial directions as well as tilt in any random direction. Eccentricity and unbalanced mass of the spindle become other sources of vibration. Inhomogeneity of the work-piece material causes 'material-induced vibration'. Additionally, external vibrations are transmitted to the cutting interface through the tool and spindle. All of these vibrations acting either individually or collectively result in hampered surface quality [13,14]. Figure 2.5 shows the effect of vibration on the surface. Since complete elimination of vibration is not possible, the approach should be to minimise the vibration transmitted to the machining zone.

2.4 Characteristics and Capabilities of Diamond Turn Machines

A characteristic of a diamond turn machine is its behaviour and capability is the output from this behaviour. Due to certain inherent characteristics of a diamond turn machine, it is possible to achieve specific capabilities. Figure 2.6 shows some of the characteristics and capabilities of diamond turn machines.

Small movement between tool and work-piece enables DTM to achieve very small unit removal (UR) of material and thus it enables exercising a very high level of control on the material removal process. Compared to a

Characteristic		Capability
Ability to achieve small movements	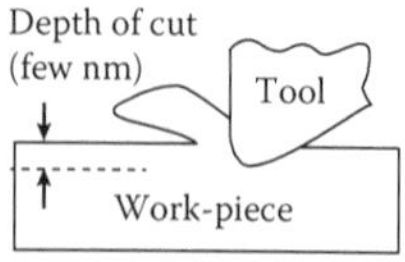 Depth of cut (few nm) Tool Work-piece	Small unit removal of material (UR) and hence better control on the process Simultaneous machining and finishing Nano-regime machining
Deterministic path control		Machining of precise surfaces and complex three dimensional geometries
Reduced vibration		Optical surface machining capability

FIGURE 2.6
Characteristics and capabilities of diamond turn machines.

precision machining process which typically employs a chip cross-section of 10 μm × 10 μm, a chip cross-section in diamond turn machining is less than 1 μm × 1 μm. This amounts to one hundredth of the UR of the precision machining process. Small UR of material in diamond turn machining facilitates simultaneous machining and finishing of surfaces.

Owing to its ability to control the machining path in a deterministic manner, it is possible to achieve nano-regime machining, i.e. the size tolerance, shape error and surface finish values are controlled within a few tens of nm. However, excessive tool wear due to small UR is one of the major disadvantages of DTM.

Minimising of all possible vibrations from external as well as from internal sources enables diamond turn machines to generate surfaces with optical quality; signatures of the vibrations like chatter and dig marks on the machined surfaces are also minimised. Damping characteristics of the bearing elements play a major role in minimising the amplitude of vibration to a greater extent.

2.5 Components of Diamond Turn Machines

Like any other machine, a diamond turn machine is built by integrating many functional components. The major components/sub-systems are shown in Figure 2.7 and the functional elements and their corresponding components are listed in Table 2.1.

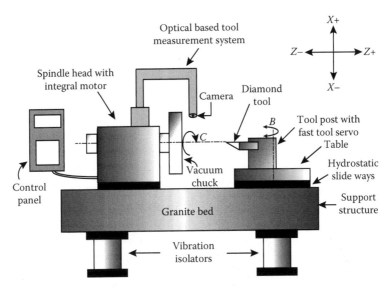

FIGURE 2.7
Components of diamond turn machine.

TABLE 2.1

Important Functional Elements and Components of a Diamond Turn Machine

	Functional Elements	Components
1	Structure and bed	Machine frame
		Enclosure
2	Positioning	Table and guide-ways
		Headstock and spindle
		Motor and drive elements
		Tool post and tool
		In-situ tool measurement
3	Control unit	Controller
4	Sensing	Linear scales and rotary encoder
		Hall sensor
		Velocity and acceleration sensors
		Closed loop feed back
5	Cooling	Spindle cooling
		Coolant for machining
6	Peripheral	Pneumatic, hydraulic and electrical devices
7	Special devices	Slow tool servo system
		Fast tool servo system

Some of the basic differences between the components of a diamond turn machine and other machines are:

- Individual components are made to achieve extremely high positional accuracy. For example, the quality of guide-ways in terms of straightness is better than a few hundred nm in diamond turn machines, compared to few μm in conventional precision machines.
- Moving components have thermal equilibrium so as to avoid differential coefficient of thermal expansion leading to inaccuracies.
- Assembly of the machine components leads towards a high level of accuracy. For example, in a headstock assembly, the spindle run-out is of the order of 20–30 nm.
- Drive systems are extremely fast responding.
- Selection of material for each component is done with considerations for inertia, thermal, damping and strength aspects.

2.6 Technologies Involved in Diamond Turn Machine Building

Building of diamond turn machines involves multiple technologies [15]. Application of such technologies enables achieving various desired functionalities in the machine and filtering out undesirable noises entering the system. For example, vibration isolator between the machine bed and the ground enables filtering out or stopping most of the incoming vibrations into the system. Similarly, use of either hydrostatic or aerostatic spindle bearings helps in achieving very high accuracy level/run-out to the tune of a few tens of nm. Table 2.2 lists various technologies used in diamond turn machine building at various component/subassembly levels. New technological developments are continuously utilised in building diamond turn machines. For example, earlier diamond turn machines used ball screw arrangements for table movement, whereas, present generation machines use linear motors for the same purpose.

Understanding the characteristics of each technology is important while applying them for building diamond turn machines. For example, in the case of table guide-ways, either hydrostatic or aerostatic bearings find applications. However, from the damping point of view, hydrostatic bearings are better suited for guide-ways, whereas from the thermal stabilisation point of view, aerostatic bearings are best suited for the spindle.

TABLE 2.2

Various Technologies Employed in Diamond Turn Machine Building

	Component/Subassembly	Technology
1	Ground–machine interface	Vibration isolator
2	Bed	Synthetic granite
3	Guide-ways	Hydrostatic bearing
4	Spindle	Aerostatic bearing
5	Tool post	Flexural mechanism
6	Compensation of slide errors and non-symmetric features	Slow tool servo [16] Fast tool servo
7	Tool position measurement	Optical/LVDT
8	Fixture	Vacuum chuck
9	Coolant	Mist
10	Spindle temperature stabilisation	Water cooling
11	Scale (table positioning)	Linear scale
12	Spindle drive	Integral brushless motor
13	Table drives	AC linear motors
14	Spindle position monitoring	Halls sensor
15	Environment	Clean room with temperature and humidity control

2.7 Environmental Requirements for Diamond Turn Machines

Environmental conditions severely affect the accuracy, performance and life of the machine. Diamond turn machines are very sensitive for the following environmental conditions:

- Temperature and humidity
- Vibration
  - Ground vibration
  - Acoustic
- Dust

In order to overcome the effects of the above factors, diamond turn machines are housed in a clean room with temperature and humidity control as well as a dust-free environment. Temperature control in the order of ±1°C and humidity control of ±5% and clean room class of 10,000 or better are generally recommended as minimum requirements to overcome the effects of temperature, humidity and dust.

As the components are machined to nanometric accuracy for size toler-ance, shape and surface finish, any variation between the temperature of the environment and work material affects the accuracy significantly. Hence, work-pieces in general are thermally stabilised by keeping them in the same environmental conditions before machining.

Transmission of ground vibration to the diamond turn machine is mini-mised by vibration isolators. However, vibrations are directly transmitted through acoustics to the machine elements and subsequently to the machin-ing zone. Hence, noise levels near the machine need to be controlled properly.

2.8 Sample Machine Specification Sheet

Table 2.3 shows sample specification of a diamond turn machine.

2.9 Summary

This chapter gives an overview about diamond turn machines. Classification of DTMs, functional requirements from DTMs including positional accuracy and repeatability as well as balanced loop stiffness, thermal and vibration effects are presented. Desirable characteristics and capabilities, various com-ponents, technologies involved in building DTM and environmental require-ments are also discussed briefly. A sample specification sheet for DTM is also provided to have a better understanding.

2.10 Sample Solved Problems

Example 1. In the diamond turn machining process, calculate the extent of inaccuracy while machining a cylinder using the following parametres:

Tool shank cross-section = 10 mm × 10 mm, length = 50 mm, material = tool steel, Young's modulus of elasticity = 200 Gpa.

Radial force (Fr) = 0.2 N

Tangential force (Ft) = 0.15 N

Stiffness of machine tool (K_m) = 50 N/μm

TABLE 2.3

Sample Specification of a Diamond Turn Machine

Feature	Description
Machine type	Diamond turn machine
Machine base	Natural/synthetic granite bed
Vibration isolation	Pneumatic, weight carrying capacity up to 5000 kg, resonant frequency in both vertical and horizontal direction < 2.5 Hz
Control system	Aerotech Model: A3200 (indicative only)
Operating system	Windows 7
Programing resolution	0.01 nm (linear)/0.0000001^0 (rotary)
Number of axes	2/3/4 X, Z, C, B
Travel	200 mm/200 mm
Swing over bed	300 mm
Resolution	16 picometer (linear)/0.018 arc-sec (rotary)
Spindle	
Bearing type	Aero static, liquid-cooled
Drive system	Integral brushless motor
Maximum spindle speed	16,000 rpm
Load carrying capacity	50 kg
Positional accuracy	±1 arc-sec
Axial stiffness	185 N/micron
Radial stiffness	75 N/micron
Table	
Bearing type	Hydrostatic
Drive system	AC linear motor
Position feedback resolution	32 picometres
Axis straightness	Horizontal: 0.2 μm, vertical: 0.3 μm
Feed rate	3000 mm/min
Achievable	
Surface finish, Ra	Better than 1.0 nm Ra
Form accuracy (P-V)	Better than 0.1 micron
Power requirement	
Electrical	Phase- 440 V, 5 KVA
Pneumatic	15SCFM @120 PSIG
Floor space (W × L × H)	1500 mm × 2000 mm × 2000 mm

Solution 1. Equivalent stiffness of the cutting zone can be calculated by considering the stiffness of the tool as well as the machine, as illustrated in Figure 2.8.

Second moment of inertia of tool shank $(I) = bd^3/12 = 10^{-2} \times (10^{-2})^3/12 = 8.3 \times 10^{-10}$ m^4.

Stiffness of tool shank $(K_t) = W/\delta_t = 3EI/L^3 = 3 \times 200 \times 10^9 \times 8.3 \times 10^{-10}/(0.05^3) = 398.4 \times 10^4$ N/m.

Or, $K_t = 3.984$ N/μm.

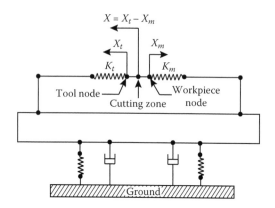

FIGURE 2.8
Loop stiffness.

Equivalent stiffness (K_{eq}) can be calculated by using the figure and force equilibrium, which is written as follows:

$$F = K_{eq}X = K_m X_m = K_t X_t => X_t/X_m = K_m/K_t$$

Note: X is tool displacement with respect to the work-piece.

$$K_{eq} = K_m X_m/X = K_m X_m/(X_t - X_m)$$

or,

$$K_{eq} = K_t X_t/X = K_t X_t/(X_t - X_m)$$

Therefore,

$$K_{eq} = K_m/(X_t/X_m - 1) = K_m/(K_m/K_t - 1) = K_t K_m/(K_m - K_t)$$

$$K_{eq} = K_t K_m/(K_m - K_t) => K_{eq} = 3.984 \times 50/(50 - 3.984)$$
$$= 199.2/46.016 = 4.33 \text{ N/}\mu\text{m}$$

Deflection due to radial force on tool $(\delta) = Fr/K_{eq} = 0.2/4.33 = 0.046 \,\mu\text{m} = 46 \text{ nm}$. This deflection is a tool path deviation with respect to the defined path program.

Thus, the extent of machining inaccuracy can be quantified as 46 nm.

Example 2. During diamond turning of copper alloy, compare inaccuracies, when cutting edge sharpness (R) varies from 100 nm to 150 nm. Assume other parametres are constant.

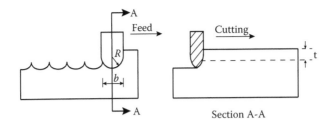

FIGURE 2.9
Ploughing force.

Solution 2. Let equivalent stiffness of tool-work-piece = K_{eq}.

Referring to Figure 2.9, ploughing force in case of 100 nm sharp tool = $F1$ = $\sigma_f b (2Rt)^{0.5} = \sigma_f b (200t)^{0.5}$ where, σ_f is flow stress.

Accordingly, ploughing force in the case of 150 nm sharp tool, $F2 = \sigma_f b (300t)^{0.5}$.

The ratio of inaccuracies due to tool deflection when cutting edge sharpness varies from 100 to 150 nm = $F1/F2$;

$$F1/F2 = (1/2)^{0.5} = 1/1.141;$$

Hence, a 150 nm sharp tool leads 1.41 times higher inaccuracy, when it is compared with 100 nm sharp tool.

2.11 Sample Unsolved Problems

Q1. During a single point diamond turn machining process, tool offset is maintained with an error value of 10 μm. Calculate form accuracy, while generating a hemispherical ball of 20 mm diameter. Assume other sources of errors are negligible.

Q2. During diamond turn machining, compare inaccuracies when it is applied on aluminium (flow stress = 140 Mpa) and brass (flow stress = 300 Mpa). Assume other parametres are constant.

Q3. In diamond turn machining, calculate the extent of vibration isolation using the following parametres for generating smooth cylindrical surfaces: tool shank cross-section = 10 mm × 10 mm, length = 50 mm, material = tool steel, Young's modulus of elasticity = 200 Gpa. Radial force (Fr) = 0.2 N, tangential force (Ft) = 0.15 N and stiffness of machine tool (K_m) = 50 N/μm.

3

Mechanism of Material Removal

3.1 Introduction

In the engineering domain, processing of material to the desired shape, size, surface finish and many other attributes is achieved either by a top-down approach or by a bottom-up approach or by a combination of both. In the top-down approach, in general, machining is carried out on bulk material using different types of tools like cutters, laser beams, electron beams etc. to remove the excess material. In the bottom-up approach, material is added to build the product by using processes like electroforming, additive manufacturing processes etc. In some cases, both approaches are combined like forging to achieve near net shape and subsequent machining to achieve the final requirements. Both top-down and bottom-up approaches have their own advantages and disadvantages. Surfaces are generated in any machining processes after tool–work-piece material interaction at every point on the generated surface; and generally, residual tensile stresses are left on the resulting machined surfaces. Whereas in most of the bottom-up processes like electroforming, building of the product takes place all over the surface simultaneously, but homogeneity of the product is not ensured. Since the method of removing or adding the material affects the quality of the product: including strength, homogeneity, surface integrity, accuracy etc., it is important to understand the mechanism of material removal or addition in product building. This knowledge will permit the user of the process to control or manipulate the process variables to minimise the undesirable effects and maximise the benefits of the specific process. Diamond turn machining (DTM), being a material removal process, has both advantages and disadvantages, which at times affect the product quality while meeting different functionalities. Even though the process parametres of DTM appear to be same as that of conventional machining, like turning and milling, owing to the variation in the cutting mechanism, many seemingly insignificant parametres of conventional machining parametres become very significant in DTM. Keeping these in view, a few important aspects of the mechanism of material removal for the following are discussed in this chapter:

- Deterministic and random processes
- Brittle and ductile materials
- Micro- and nano-regime cutting

3.2 Comparison of Deterministic and Random Machining Process

DTM removes a very small amount of material from the preshaped component and hence it can be considered either a machining or a finishing process. Therefore, machining and finishing are used in an interchangeable manner in this chapter. Among other things, tool and work-piece interactions in either tool- or abrasive-based machining process significantly affect the quality of the finished surface. Control on the quality factors of the finished surface like shape, size and surface finish are dictated by the process being employed. Therefore, it becomes imperative to understand the characteristics, behaviour of the finishing process and their capability to generate the desired quality of surface. Based on their characteristics, the finishing processes can be broadly classified as

- Deterministic finishing processes
- Random finishing processes

Figures 3.1 and 3.2 schematically show typical examples for these two classes of processes. A typical deterministic process, namely, the turning process, is shown in Figure 3.1. In this process, the unit removal (UR)

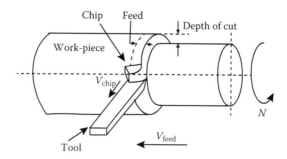

FIGURE 3.1
Schematic of deterministic finishing process: Turning.

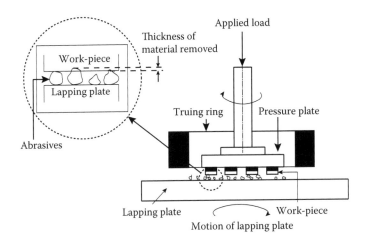

FIGURE 3.2
Schematic of random finishing process: Lapping.

of material can be precisely controlled by way of controlling the process parametres. In this case, UR is the function of depth of cut, feed/rev and cutting velocity. Additionally, controlling the tool path enables machining of any complex surfaces. This ability to control the tool path travel (on the surface being generated) in a deterministic manner with predictable material removal rate enables achieving the desired size and shape of the component in a faster manner. Typical material removal rate (MRR) for the turning process is expressed by the following equation:

$$MRR = d\,f\,v\,\text{mm}^3/\text{min} \qquad (3.1)$$

where d = depth of cut in mm; f = feed in mm/revolution; v = cutting velocity in mm/min.

A typical random process, namely, the lapping process, is shown in Figure 3.2. In this process, the extent of material removal from the surface varies from location to location. However, it takes place simultaneously at different tool – work-piece (material) contact locations. Therefore, in this type of process, prediction of precise material removal rate is difficult and varies in a nonlinear fashion. In this process, tool path control on the work surface is not possible and hence this process is limited to finish only simple geometries. Typical average MRR for the flat lapping process is expressed by Preston's equation:

$$Average\ MRR \propto Pv = dT/dt \qquad (3.2)$$

where P = pressure, v = Volume; T = thickness in mm; and t = time in min.

TABLE 3.1

Key Characteristics and Capabilities of Deterministic and Random Finishing
Processes

	Deterministic Process	Random Process
Characteristics	Path controlled	Force controlled
	Location of point of material removal is controllable	Material removal by area averaging
	Few process parametres	Large number of process parametres
	Faster process	Slow and tedious process
	Machine needs to be precise, as its signature is transferred to generated surface	Machine need not to be precise, as its signature is not transferred to generated surface
	Lay pattern is generated on finished surface	No lay pattern is generated on finished surface
Capabilities	Precisely controllable MRR	Difficult to control MRR
	Complex surface generation is possible	Only generation of simple surfaces like flat and spherical surfaces are possible
	Size control is possible	Size control is not possible
	Finish impaired by lay pattern	Extremely high surface finish

Due to its area averaging capability and ability to control the cutting
force, the flatness and surface finish (achievable in the lapping process) are
extremely high, whereas the deterministic processes leave the lay pattern on
the generated surface. Another major difference between these processes is
that, machine inaccuracies are reflected on deterministically finished surfaces, whereas, the machine inaccuracies are not reflected on random finished surfaces.

As mentioned earlier, both processes have their own advantages and disadvantages. Table 3.1 lists some of the key characteristics and capabilities of
both deterministic and random finishing processes. DTM, which is the focus
of the discussion, falls under the deterministic finishing process.

3.3 Cutting Mechanisms for Engineering Materials

Engineering materials are broadly classified as ductile and brittle materials. The failure modes in these materials help us to understand the
mechanism of material removal. Since one of the major objectives of
machining is to remove material from the work surface and subsequently
achieve desired qualities on the generated surface, it becomes important
to understand the mechanism of material removal in various engineering

materials. Material removal mechanism is generally affected by the following factors:

- Work material and its properties
- Tool material and its properties
- Relative position between the work-piece surface and tool
- Relative motion between the work material and tool

Figure 3.3 depicts the process of material removal in ductile materials. Numerous factors play a significant role in the material removal mechanism and subsequently affect the quality of the generated surface.

In the case of ductile material, when the material is compressed by the moving tool, it slides along the shear zone as indicated in Figure 3.3, gets converted into 'chip' due to plastic deformation and subsequently is removed from the parent material. Depending on the defect density of the shear zone, the specific cutting energy necessary to remove the material varies. As the uncut chip thickness reduces, the defect density in the shear zone also reduces, and this leads to the requirement of higher specific cutting energy to remove material as chip. During the material cutting process, the desired component shape is achieved by controlling the tool path. Table 3.2 summarises some of the factors affecting the ductile material removal process.

The specific cutting energy in machining ductile materials can be estimated by the following equation:

$$\text{Specific Cutting Energy} = G\,e^{(-2\pi W/a)} \tag{3.3}$$

where G = modulus of rigidity, W = dislocation width and a = interatomic spacing.

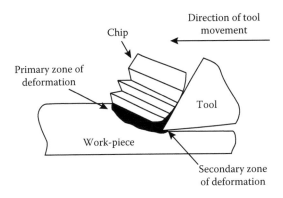

FIGURE 3.3
Ductile material machining mechanism.

TABLE 3.2

Factors Affecting Ductile Material Removal

Source	Property	Effect
Material	Ductility Work hardening Grain size	Material flow back characteristic and rubbing of finished surface with tool
Cutting tool	Cutting edge radius	Minimum achievable uncut chip thickness
	Cutting tool angles/geometries/ orientations Hot hardness	Tool rubbing on the finished surface
Machining parametres	Uncut chip thickness	Rubbing/ploughing/cutting
Machine	Precision	Minimum uncut chip thickness Transfer of machine signature on finished surface

For a given material, the most important factor affecting the cutting mechanism is the tool nose radius. For a given tool edge radius r, the mechanism changes from cutting by shear to ploughing to rubbing, when the uncut chip thickness (t_c) changes from above the threshold value of uncut chip thickness value to below the threshold value. Figure 3.4 shows the same schematically and Table 3.3 shows the conditions of material removal mechanism for a variety of cutting edge sharpness.

For a given work material and cutting edge radius of the tool, when the uncut chip thickness is greater than critical chip thickness (t_c), shearing by plastic deformation becomes predominant and chip is formed from the displaced work material [17] as shown in Figure 3.4b. When the uncut chip thickness is between critical chip thickness and the thickness that causes pure elastic deformation, elastic deformation becomes predominant and ploughing action takes place as shown in Figure 3.4c and no material is removed from the work-piece. When the uncut chip thickness is less than the value corresponding to pure elastic deformation, rubbing of tool on the work surface takes place and no material is removed from the work-piece as shown in Figure 3.4d. At certain critical values of uncut chip thickness, the rake angle of the tool becomes positive to negative and it creates a dead metal zone leading to increasing cutting force. This increases the specific cutting force and specific cutting energy. This phenomenon of increasing specific cutting force or specific cutting energy with decreasing uncut chip thickness is known as size effect. Figure 3.4e shows the change in the mechanism with a/r ratio; when the ratio of a/r decreases for a given uncut chip thickness, the mechanism of material removal changes from shearing to ploughing and subsequently to rubbing.

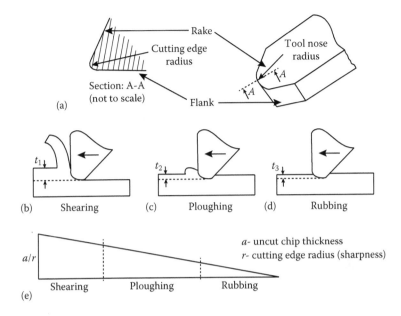

FIGURE 3.4
Effect of tool nose radius on cutting mechanism. (a) Tool nomenclature. (b) Shearing. (c) Ploughing. (d) Rubbing. (e) Cutting mechanism for different a/r ratio.

TABLE 3.3

Effect of Cutting Edge Radius on Cutting Mechanism

		Increasing cutting edge radius $\longrightarrow$		
		r_1	r_2	r_3
Decreasing uncut chip thickness	t_1	Cutting	Ploughing	Rubbing
	t_2	Ploughing	Rubbing	–
	t_3	Rubbing	–	–

Note: where $r_1 < r_2 < r_3$, $t_1 > t_2 > t_3$

Figure 3.5 explains the cutting mechanism for brittle material machining. When the indenter representing the cutting tool exerts pressure on the material (here it is shown normal to the machining surface), after certain penetration the material generates lateral cracks and subsequently median cracks in the region of indentation as shown in Figure 3.5. When these lateral and median cracks meet, material removal takes place. This leads to generation of discontinuous chips in brittle materials [18,25]. Figure 3.6 schematically shows the formation of discontinuous chips in brittle material processing.

Maximum theoretical specific cutting energy in brittle material machining is given by the following equation:

$$\text{Specific cutting energy} = G/2\pi. \tag{3.4}$$

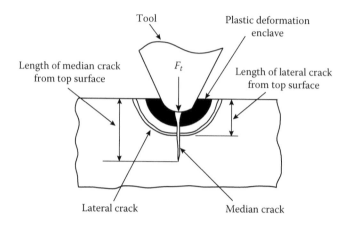

FIGURE 3.5
Brittle material machining.

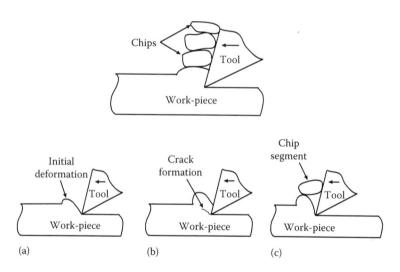

FIGURE 3.6
Formation of discontinuous chips in brittle material processing. (a) Initial deformation of material. (b) Crack formation. (c) Chip formation.

In general, material removal in both ductile and brittle materials can be schematically represented by the multigrain material removal process as shown in Figure 3.7. In multigrain machining, which takes place at the macro level, material is removed due to the failure line moving along the grain boundaries as shown in Figure 3.7. The resistance offered to the plastic deformation of the material is nearly equal to the shear strength of the material. In such cases, when the grain size becomes smaller, the number of grain boundaries increases and results in increased resistance for material removal.

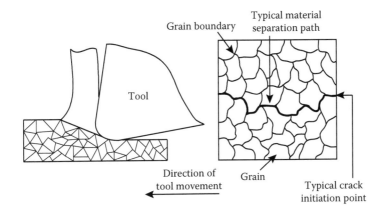

FIGURE 3.7
Multigrain mode material removal.

3.4 Micro- and Nano-Regime Cutting Mechanisms

When the uncut chip thickness gradually decreases, removal of the material takes place across the grains as shown in Figure 3.8. In this case, the typical failure path shown in the figure moves across the grain instead of, along the grain boundary [18]. Resistance offered for the material removal becomes enormous, as the defect density along the path of failure becomes much lesser. Defects like dislocations and vacancies within a grain play a major role in providing the failure path. As the uncut chip thickness value reduces from the micro to the nano level, the defect density becomes lesser and the resistance to the removal of the material increases significantly [19,20]. This resistance to the material removal also affects the tool life and

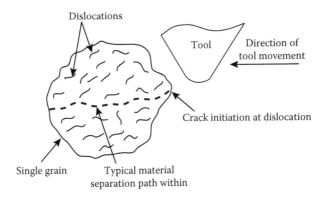

FIGURE 3.8
Sub-grain material removal.

causes its deterioration. Figure 3.9 shows a typical graph indicating the correlation between uncut chip thickness and the specific cutting energy. At the atomic level of uncut chip thickness, the specific cutting energy is equivalent to the atomic bonding force.

Figure 3.10 shows various regions for a typical cutting process. Regions I, II and III represent the regions corresponding to nano cutting, micro cutting and macro cutting, respectively.

Unlike the single point machining process, the abrasive-based micro- and nano-finishing processes gradually remove only the projecting peaks, until all of them are smoothed out. Once all the peaks are removed, material removal slows down. Figure 3.11 shows various stages of material removal by a single loosely held abrasive. Figure 3.11a shows the interaction of abrasive particle with the peak of the micro irregularity, and Figure 3.11b shows a partially sheared off chip over the peak. In this stage, the abrasive particle and chip are joined due to the secondary bonding force. As the length of shear plane along the cutting path gradually reduces, the material resisting force becomes lesser than the secondary bonding force existing between the abrasive and chip. Subsequently, it leads to the removal of the chip.

As experimental techniques cannot be applied effectively when uncut chip thickness is at the nanometric level, the molecular dynamics simulation (MDS) technique is extensively used to explain the material removal behaviour at the nanometer scale level [21,22]. Figure 3.12a shows a typical molecular dynamic simulation for a tool–work-piece interaction. There are three types of atomic layer regions, namely, the Newtonian layer, the thermostat layer and the boundary layer. The Newtonian layer atoms follow Newton's second law of motion under specified pair

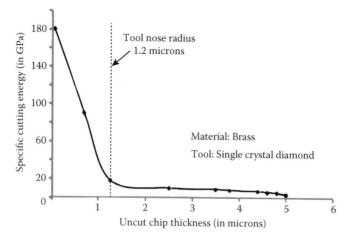

FIGURE 3.9
Effect of uncut chip thickness on specific shear energy.

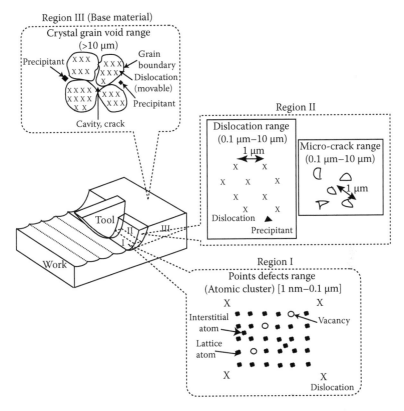

FIGURE 3.10
Macro, micro, and nano cutting regions.

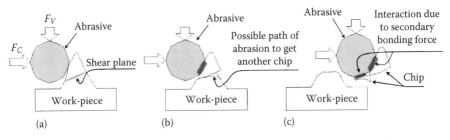

F_C - Cutting force F_V - Holding force

FIGURE 3.11
Material removal by abrasive particle. (a) Abrasive starts touching the work-piece. (b) Abrasive shearing the peak. (c) Peak sheared as chip.

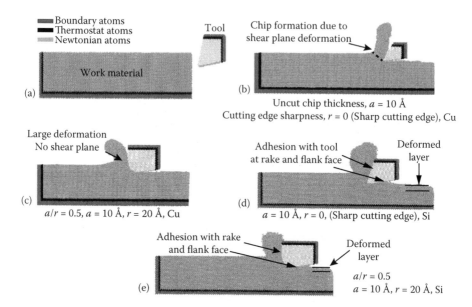

FIGURE 3.12
(a) MDS layers; (b) MDS for Cu with sharp tool; (c) MDS for Cu with finite tool sharpness; (d) MDS for Si with sharp tool and (e) MDS for Si with finite tool sharpness.

potential interaction. Whenever two atoms come near to each other, they are subjected to inter-atomic force which is a negative derivative of the pair potential. This force is used to move the atoms. Velocity and position of the atoms are computed using the Verlet algorithm. Thermodynamic aspects are computed by applying conservation of number of atoms, volume and energy of the Newtonian region. During atomic movements, a rise in the local temperature is carried away by the thermostat layer. All atoms in this region will have the same temperature. Atoms in the boundary layer region provide fixed boundary condition. Among other factors, MDS helps to visualise and analyse the following:

- Shear plane and chip formation
- Type of crystal
- Temperature
- Phase transformation
- Interaction between two bodies like work material and tool

Figure 3.12b and c show MDS for different ratios of uncut chip thickness to tool edge radius, when copper and single crystal diamonds are used as work material and tool material, respectively. Similarly, Figure 3.12d and e show MDS for different ratios of uncut chip thickness to tool edge radius or sharpness, when silicon (Si) and single crystal diamonds are used as work material and tool material, respectively.

3.5 Ductile Regime Machining of Brittle Materials

Under certain cutting conditions, brittle materials behave like ductile materials. When the uncut chip thickness is less than a certain threshold value $[t_c]$, which varies for different materials, the brittle material ahead of the cutting tool is removed as a continuous chip in plastic deformation mode. When the defects initiated ahead of the tool do not penetrate into the finished surface, the generated machined surface of the brittle material is free from surface defects. Such a condition is achieved when the uncut chip thickness is less than t_c. Blake and Scattergood [23] presented this critical-depth parametre, which governs the transition from plastic flow to fracture along the cutting edge nose radius. They have reported that owing to complex interaction between tool geometry, machining parametres and material response, a major portion of material removal occurs by fracture even when ductile-regime conditions are achieved [23,24]. Typical values of t_c for different materials are shown in Table 3.4. Figure 3.12d shows MDS for ductile regime machining of Si, when machined with uncut chip thickness less than t_c.

In case of brittle materials, threshold chip thickness is expressed by the following equation [25]:

$$t_c = a \times \frac{E}{H} \times \left(\frac{K_c}{H} \right)^2 \tag{3.5}$$

TABLE 3.4

Threshold Uncut Chip Thickness for Different Brittle Materials

Material	Threshold Uncut Chip Thickness (t_c) in nm	Reference
Si	200	[26]
	236	[27]
BK7 Glass	62	[27]
Ceramics:		
PC5K	0.043	[28]
PC4D	0.170	
Nano crystalline, binderless tungsten carbide	165	[29]

where a = constant depends on materials; E = elastic modulus; K_c = fracture toughness of the material; and H = hardness of the material.

3.6 Machining of Polymers

Precision polymer optics are extensively used in many fields. As the accuracy achievable by injection moulding, compression moulding or extrusion of polymers is inadequate to meet the requirements for many optical applications, precision machining by tool-based machining processes becomes necessary. Most of the polymers are soft, not as strong as metals, less rigid, have low density, thermally insulating and are viscous under elevated temperatures.

Figure 3.13 shows the structure of typical amorphous polymers. In one type, the polymer chains are randomly distributed and in another type very long polymer chains are arranged in regular fashion.

In the case of metals and ceramics, due to their high rigidity, distortion on the machined surface is resisted by the material. However, in the case of polymers, they offer lesser resistance for cutting force and the machined surface is distorted significantly. Due to the following properties of the polymer, the generated surface is significantly affected by machining [30]:

- Glass transition temperature (T_g)
- Viscoelastic property
- Relaxation time

The glass transition temperature (T_g) is the temperature region where the polymer transition from a hard material to soft material takes place drastically and it is always less than the melting temperature.

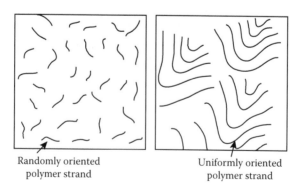

Randomly oriented   Uniformly oriented
polymer strand   polymer strand

FIGURE 3.13
Arrangement of polymer chains.

Viscoelasticity is the property of polymer that exhibits both viscous and elastic characteristics when undergoing deformation. Due to the viscoelasticity property of polymers, deformation under stress becomes time dependent. When the imposed mechanical stress is held constant, the resultant strain increases with time and when constant deformation is maintained with loading, the induced stress relaxes with time.

At a given temperature, every polymer has a fixed response time, which is related to the sum of its vibrational, rotational and translational movements. This characteristic time is defined as relaxation time. The relaxation time decreases with increasing temperature and the polymer becomes softer and more ductile with escalating temperature.

During the machining process, polymers exhibit the following three distinct behaviours:

- Glassy state
- Elastic or rubber like state
- Liquid or fluid flow state

The machining of polymers is similar to other engineering materials, except that the temperature rise causes the material removal process to behave like brittle fractures at the initial phase, when the temperature is below T_g. It behaves like ductile shear plastic deformation once it crosses T_g. This temperature depends on the type of polymer and degree of crosslinking of the polymer strands [31]. Monomers are held together either by covalent bond or by ionic bonds to form polymers. Many polymer strands are held together by either ionic bonds or by weak Van der waal's forces. This force can be derived from Lenard–Jones potential [30], given by

$$V_{ij} = 4\varepsilon \left[\left(\frac{\sigma}{r} \right)^{12} - \left(\frac{\sigma}{r} \right)^{6} \right] \tag{3.6}$$

where r is inter-atomic spacing and σ and ε are constants that are dependent on the physical property of the materials. During the material removal process, sliding of polymer strands with respect to each other as well as by shearing or by cutting of strands takes place.

Major problems encountered during polymer precision machining include

- Poor chip disposal
- Reattachment of fine chips on the finished surface
- Melting of the finished surface
- Dimensional distortion of the component due to mechanical and thermal effects

3.7 Summary

This chapter discusses various aspects of material removal mechanisms both in macro and micro domains. Different types of finishing methods and mechanisms of material removal for various types of engineering materials including ductile, brittle and polymers are discussed. The mechanisms of material removal in diamond turn machining including micro- and nano-regime machining and molecular dynamic simulations are also briefly discussed.

3.8 Sample Solved Problems

Example 1. During diamond turn machining of a copper alloy, calculate the 'specific cutting energy' using the following parametres. Rigidity of copper alloy (G) = 45 GPa; Poisson's ratio (v) = 0.28, inter-planer spacing (d_{hkl}) = 0.288 nm, inter-atomic spacing (a) = 0.2556 nm, cutting edge sharpness of the diamond tool = 5.0 μm.

Solution 1. Here, the cutting edge sharpness of the tool is of the order of a few microns (i.e. 5 μm). In this case, chip deformation depends on the dislocation mechanism and hence 'Peierls stress' shall be used to compute flow stress for the chip formation. Dislocation width (W) = $d_{hkl}/(1 - v)$ = 0.288/(1 − 0.28) = 0.4 nm.

Shear stress to move dislocation (τ_p) can be calculated using 'Peierls stress' formulation for the chip 'flow stress' as follows:

$$\tau_p = G\,e^{(-2\pi W/a)} = 45 \exp\left[(-44/7)\times(0.4/0.2556)\right] = 131.2\times 10^{-3}\ \mathrm{GPa} = 131.2\ \mathrm{MPa}.$$

Or, τ_p = 131.2 Mpa. Thus, the required specific cutting energy (Υ) = energy to deform (or strain energy due to shear stress).

$$\Upsilon = \tau_p^2/2G = 131.2^2/90{,}000 = 0.192\ \mathrm{MPa} = 0.192\ \mathrm{J/cm^3},\ \text{where 1 MPa = 1 J/cm}^3.$$

Example 2. During diamond turn machining of a copper alloy, calculate the 'specific cutting energy' using the following parametres. Rigidity of copper alloy (G) = 45 GPa; Poisson's ratio (v) = 0.28, inter-planer spacing (d_{hkl}) = 0.288 nm, inter-atomic spacing (a) = 0.2556 nm, cutting edge sharpness of the diamond tool = 50 nm. Assume specific cutting energy at the nanometric scale is 90% of the atomic scale.

Solution 2. Required shear stress at the atomic scale (theoretical shear stress) $\tau_{max} = G/2\pi$. Therefore, shear stress to flow the material $(\tau) = 0.9\ \tau_{max} = 0.9 \times 45/2\pi$ GPa = 6.44 GPa.

Thus, the required specific cutting energy (Υ) = energy to deform (or strain energy due to shear stress):

$$\Upsilon = \tau_2/2G = 6.44^2/90 = 0.4608 \text{ GPa} = 460.8 \text{ MPa} = 460.8 \text{ J/cm}^3;$$
$$\text{where } 1 \text{ MPa} = 1 \text{ J/cm}^3$$

Example 3. During diamond turn machining of ductile (copper) and brittle (silicon) materials, compare the 'specific cutting energy' using the following parametres. Rigidity of copper (G_{Cu}) = 45 GPa; Poisson's ratio (ν) of copper = 0.28, inter-planer spacing (d_{hkl}) of copper = 0.288 nm, inter-atomic spacing (a) of copper = 0.2556 nm; rigidity of silicon (G_{Si}) =125 Gpa. Assume machining is carried out under the ductile mode of machining and generated crack-free surfaces. Cutting edge sharpness of the diamond tool = 200 nm.

Solution 3. In the case of copper, the required shear stress can be calculated using 'Peierls stress':

$$\text{Dislocation width } (W) = d_{hkl}/(1-\nu) = 0.288/(1-0.28) = 0.4 \text{ nm}$$

Shear stress to move dislocation $(\tau_{Cu}) = G_{Cu}\ e^{(-2\pi W/a)} = 45$ exp $[(-44/7) \times (0.4/0.2556)] = 45 \times 2.915 \times 10^{-3}$ GPa. Or, $\tau_{Cu} = 131.2$ MPa.

Thus, the required specific cutting energy (Υ_{Cu}) = energy to deform (or strain energy due to shear stress):

$$\Upsilon_{Cu} = \tau_{Cu}^2/2G_{Cu} = 131.2^2/90,000 = 0.192 \text{ MPa} = 0.192 \text{ J/cm}^3;$$
$$\text{where } 1 \text{ MPa} = 1 \text{ J/cm}^3$$

Thus, the required specific cutting energy in copper $(\Upsilon_{Cu}) = 0.192$ J/cm³.

In the case of silicon, the material deformation is taking place under the ductile mode of shear. Hence, silicon material will flow without generating cracks and a significant number of dislocations. Thus, the shear stress in this case shall be approximately equal to theoretical shear stress:

$$\tau_{Si} = G_{Si}/2\pi = 125/2\pi = 19.89 \text{ GPa}$$

$$\Upsilon_{Si} = \tau_{Si}^2/2G_{Si} = 19.89^2/250 = 1.58245 \text{ GPa} = 1,582.45 \text{ MPa}$$

Thus, the required specific cutting energy in silicon $(\Upsilon_{Si}) = 1{,}582.45$ J/cm³.

$$(\Upsilon_{Cu} = 0.192 \text{ J/cm}^3) \ll (\Upsilon_{Si} = 1{,}582.45 \text{ J/cm}^3)$$

Example 4. During diamond turn machining of ductile (copper) and brittle (silicon) materials, compare the 'specific cutting energy' using the following parametres. Rigidity of copper $(G_{Cu}) = 45$ GPa; Poisson's ratio (v) of copper $= 0.28$, inter-planer spacing (d_{hkl}) of copper $= 0.288$ nm, inter-atomic spacing (a) of copper $= 0.2556$ nm, rigidity of silicon $(G_{Si}) = 125$ Gpa, Poisson's ratio (v) of silicon $= 0.24$, dislocation width of silicon $= 0.25$ nm, inter-atomic spacing (a) of silicon $= 0.23$ nm. Assume machining is carried out under the ductile mode of machining and generated crack-free surfaces. Cutting edge sharpness of the diamond tool $= 5.0$ µm.

Solution 4. As cutting edge sharpness of the diamond tool $= 5.0$ µm, material deformation will take place because of dislocation movement in both cases.

In the case of copper, required shear stress can be calculated using 'Peierl's stress':

$$\text{Disclocation width } (W) = d_{hkl}/(1-v) = 0.288/(1-0.28) = 0.4 \text{ nm}$$

Shear stress to move dislocation $(\tau_{cu}) = G_{Cu} \, e^{(-2\pi W/a)} = 45 \exp [(-44/7) \times (0.4/0.2556)] = 45 \times 2.915 \times 10^{-3}$ GPa. Or, $\tau_{cu} = 131.2$ Mpa.

Thus, the required specific cutting energy $(\Upsilon_{Cu}) =$ energy to deform (or strain energy due to shear stress):

$$\Upsilon_{Cu} = \tau_{cu}^2/2G_{Cu} = 131.2^2/90{,}000 = 0.192 \text{ MPa} = 0.192 \text{ J/cm}^3$$

In the case of silicon, the material deformation is taking place under the ductile mode of shear. Hence, silicon material will flow due to dislocations, as the cutting edge radius is of the order of a few microns and without generating cracks. Thus, the shear stress in this case shall be calculated using 'Peierl's stress'.

Shear stress to move dislocation $(\tau_{Si}) = G_{Si} \, e^{(-2\pi W/a)} = 125 \exp [(-22/7) \times 0.25/0.23] = 4.1$ GPa.

$$\Upsilon_{Si} = \tau_{Si}^2/2G_{Si} = 0.06724 \text{ GPa} = 67.24 \text{ MPa} = 67.24 \text{ J/cm}^3$$

$$(\Upsilon_{Cu} = 0.192 \text{ J/cm}^3) < (\Upsilon_{Si} = 67.24 \text{ J/cm}^3)$$

3.9 Sample Unsolved Problems

Q1. During diamond turn machining of an aluminium alloy, calculate the 'specific cutting energy' using the following parametres. Rigidity of the aluminium alloy (G) = 25 GPa; Poisson's ratio (ν) = 0.26, inter-planer spacing (d_{hkl}) = 0.3 nm, inter-atomic spacing (a) = 0.25 nm, cutting edge sharpness of the diamond tool = 5.0 μm.

Q2. During diamond turn machining of an aluminium alloy, calculate the 'specific cutting energy' using the following parametres. Rigidity of aluminium alloy (G) = 25 GPa; Poisson's ratio (ν) = 0.26, inter-planer spacing (d_{hkl}) = 0.3 nm, inter-atomic spacing (a) = 0.25 nm, cutting edge sharpness of the diamond tool = 50 nm. Assume specific cutting energy at the nanometric scale is 90% of the atomic scale.

Q3. During diamond turn machining of ductile (aluminium) and brittle (silica) materials, compare the 'specific cutting energy' using the following parametres. Rigidity of aluminium (G_{Al}) = 25 GPa, Poisson's ratio (ν) of aluminium = 0.26, inter-planer spacing (d_{hkl}) of aluminium = 0.3 nm, inter-atomic spacing (a) of aluminium = 0.25 nm, rigidity of silica (G_{Silica}) = 50 Gpa. Assume machining is carried out under the ductile mode of machining and generated crack-free surfaces. Cutting edge sharpness of the diamond tool = 100 nm.

Q4. During diamond turn machining of ductile (aluminium) and brittle (silica) materials, compare the 'specific cutting energy' using the following parametres. Rigidity of aluminium (G_{Al}) = 25 GPa, Poisson's ratio (ν) of copper = 0.26, inter-planer spacing (d_{hkl}) of aluminium = 0.288 nm, inter-atomic spacing (a) of aluminium = 0.25 nm, rigidity of silica (G_{Silica}) = 50 Gpa, Poisson's ratio (ν) of silica = 0.16, dislocation width of silica = 0.25 nm, inter-atomic spacing (a) of silica = 0.22 nm. Assume machining is carried out under ductile mode of machining and generated crack-free surfaces. Cutting edge sharpness of the diamond tool = 5.0 μm.

4

Tooling for Diamond Turn Machining

4.1 Introduction

As should be evident to the reader by now, the DTM process is a very unique machining process that provides to the work material, very fine form and surface finish, often needed in optical applications. In order to provide such a finish, there are stringent requirements on the cutting tool materials and geometry to be used in the process. The surface finish obtained is directly a function of the chip formation mechanism and the nature of deformation in the work surface close to the cutting tool edge. Maintaining this mechanism uniformly throughout the path of cut and at various regions of the cutting edge is critical to keep the surface finish fine and uniform throughout. An important feature of the cutting tool that helps in this is the sharpness of the cutting edge. The tool sharpness of tens of nanometres is required in this process. Achieving such sharpness is normally not possible in polycrystalline aggregates and only single crystal materials such as diamond can meet these requirements. In addition, the form of the machined surface depends on the tool edge geometry and its uniformity around the edge. Any deviation from the planned tool geometry due to wear and poor tool fabrication will lead to deviation from the desired form, as the cutting tool traces the computer numerical control (CNC) motion paths. Hence, fabricating the tool with correct cutting edge form with controlled edge waviness is critical. This chapter discusses the need for special cutting tools to be used in the diamond turn machining process. It further provides explanation about these cutting tools, how they are fabricated, how to set them in the DTM machine and issues related to tool wear and economics of the process.

4.2 Tool Materials and Their Requirements

Just as in conventional machining, the basic requirements of a cutting tool need to be met for the DTM process. The tool should have and maintain sufficient hardness at the machining conditions involved, should be tough (resist fracture) and be resistant to gradual wear – abrasive, adhesive and chemical resistance. Based on these requirements, one can pick a plethora of cutting tools available from the conventional machining tool menu. However, these conventional cutting tools do not provide the needed surface finish conditions demanded from the DTM process. One of the main reasons for this is fabricating and maintaining the condition of the cutting edge to exacting sharpness, smoothness and uniformity. The cutting edge, ideally a line/curve but practically a surface, is the intersection of the rake face and the clearance faces of the cutting tool. The cutting edge normally is characterized by two radii – the nose radius and the edge radius (Figure 4.1). The basic requirement for a cutting tool to perform well in DTM is that, the, cutting edge surface has to have fine surface finish and precision form tolerance – both along the edge and across the edge. The curved cutting edge surface spans several millimetres along its length, but is only a few tens of nanometres across its width. Both forms – along the length and across – are typically circular in shape and characterized by a radius – nose and edge radii as mentioned earlier. The form tolerance along the edge is usually reported as waviness, while the sharpness which is formed across the edge is not normally controlled or any specification laid out explicitly. Both the waviness (deviation from circularity along the length) and edge radii are in the range of a few hundreds of nanometres to a few tens of nanometres respectively and pose a major challenge in the fabrication of a cutting tool for DTM applications.

Most of the commercially successful cutting tool materials for conventional machining (carbides, poly crystalline diamond [PCD], polycrystalline cubic boron nitride [PCBN], ceramics) are aggregates of smaller particles cemented together using a binder. Such aggregates, due to their random orientation of the individual particulates, are fairly isotropic in their properties. As shown in Figure 4.2b, the microstructure of a poly-granular or poly-crystalline

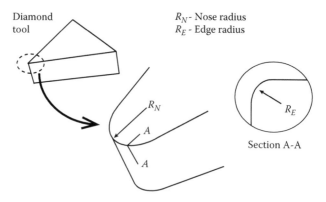

FIGURE 4.1
Schematic explaining nose radius and edge radii in a cutting tool.

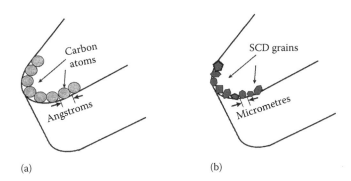

FIGURE 4.2
Schematic explaining the difference between PCD and SCD. (a) Single crystal tool and (b) poly crystalline tool.

aggregate, however, prevents the fabrication of a smooth cutting edge – the granular aggregates naturally form protrusions at the two meeting faces, which cannot be smoothened out effectively enough at the nanometric scale.

One solution to this problem is to use single crystal materials to form the cutting edge – the only edge protrusions then possible, theoretically, are the atomic spherical protrusions (Figure 4.2a). However, single crystals (SC) are notoriously anisotropic in their properties and not all SC materials can be easily grown into large sizes suitable for cutting tools. Materials such as carbides, ceramics and CBN can only be grown to single-crystal sizes of several tens/hundreds of micrometres and not any bigger. The only material that can be artificially grown to, or is available naturally in, large sizes and that has reasonable properties in many crystal directions is diamond. It also has excellent hardness and abrasive wear resistance (although varying in different crystal planes and directions). Given that the SC material is not isotropic and has varying properties along its crystallographic planes and directions, choosing the best one (and how to do this) for fabricating the cutting edge surface is also a major challenge to overcome. Such problems do not occur in the fabrication of cutting tools used in conventional machining.

4.3 Single Crystal Diamond Tools

The one single crystal material that has largely been successful as a cutting tool in the DTM process is diamond. Both artificially synthesised and naturally available diamonds have been used for fabricating cutting tools with some users indicating preference for natural diamonds. Commercial single crystal diamond (SCD) tools are available for users to buy and companies using the

DTM process rely on such suppliers (e.g. UK-based Contour Diamond, Japan-based ALMT Diamond Corporation, etc.) for their regular use. Relapping of used/worn SCD tools are also carried out by these companies with SCD tool inserts going through multiple relaps before being discarded.

Diamonds used for cutting tools in DTM are obtained from naturally available sources or synthetically made. Synthetic diamonds are made by a CVD process and are commercially available with flat surfaces of known crystallographic planes. Cutting tool manufacturers can then use these surfaces as datum to create desired geometric shapes for the tool. Natural diamonds used in cutting tools are usually aesthetic rejects from the jewelry industry and are usually available in a unfaceted raw form. However, certain defects present in natural diamonds also prevent successful performance as cutting tools. Brownish tinges in the diamond indicate the presence of residual stresses that can cause tool edges to crack and chip off during machining. Presence of carbon spots is also an indication of hole-formation tendency in the crystals. Natural crystal growth can be irregular in many places, leading to inherent property variation within the crystal. Tool manufacturers using natural diamond face such multiple challenges in picking the best diamond. Next challenge is to detect these prescribed crystallographic planes and directions inside the raw crystal; the diamond crystal then has to be fixtured (held) in such orientations during the processing steps needed to make the tool.

Single crystals have anisotropic properties owing to the discrete spatial arrangement of atoms and diamond is no exception. It is important to understand the effect of various crystallographic orientations and directions on desired properties so that, the best possible plane and direction can be chosen for the rake face, flank face and cutting edge of the tool.

The structure of diamond is the well-known tetrahedral arrangement of sp^3 hybridised carbon atoms with a central carbon forming bonds with four adjacent atoms in the tetrahedron. Four such tetrahedra are arranged in four (of the eight) alternating sub-cubic cells inside an FCC cell arrangement of C-atoms (Figure 4.3). This set of four tetrahedra form the unit cell of diamond.

In this unit cell structure, one can examine the predominant planes and directions within each plane. Not much attempt has been made to explore the most optimum crystallographic plane for cutting tool applications. However, some main planes in diamond have been extensively explored for properties such as abrasive wear, friction, etc. The predominant planes often considered are the {111}, {110} and the {100} family of planes (Figure 4.4).

The questions that come up then are the following:

- Which of these crystallographic planes should be the rake face of the cutting tool?
- Which of these crystallographic planes should be the clearance faces? Clearance often forms in multiple planes.

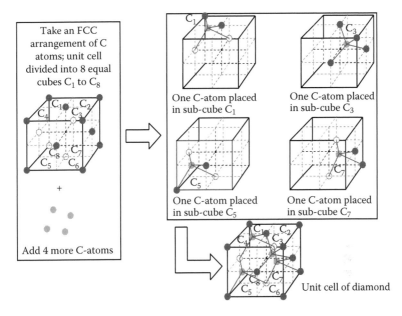

FIGURE 4.3
Schematic showing how carbon atoms are arranged in a diamond. Imagine an FCC cell of C atoms. Divide this into eight equal cubes. At the center of four cubes (C_1, C_3, C_5 and C_7 – numbered as shown), place a carbon atom and tetragonally bond it to the three nearest face centered atoms and one nearest edge atom.

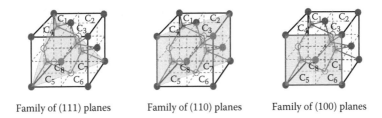

| Family of (111) planes | Family of (110) planes | Family of (100) planes |

FIGURE 4.4
Primary planes considered in diamond cubic structure.

- In which direction on the rake face plane should the cutting edge be fabricated? Cutting edge, often being nonlinear, is normally along, i.e. tangential to, multiple directions.

Two properties: abrasive wear and friction coefficient and their crystallographic dependence answer these questions. The ease with which a diamond crystal undergoes abrasive wear is the key to its successful shaping into a cutting tool. The only way to shape the diamond, being the hardest natural substance known, is using the diamond itself. Abrasive diamond powders are used to shape diamond crystals by abrasive wear (tool fabrication is described in another section in this chapter). Directions that are too hard to wear out, while ideal as cutting tool surfaces require substantial and hence uneconomic processing times. For example, the family of planes {111} is known to be the toughest plane to abrade. In this plane, the direction <110> is the toughest to process. The softest plane is the {100} family of planes; in this plane the <100> direction is the easiest to abrade. Experimental data for these observations are readily available [32]. A summary of the various softest directions to abrade in various planes is shown in Figure 4.5. The key to successful fabrication of a well-performing cutting tool is to determine and hold (fixture) the diamond crystal in the desired plane and direction.

The other property of consideration is the friction coefficient. The cutting edge should be carved out of the crystal in such a way that the chip flow direction is along the one with the lowest coefficient of friction. This is a rule that cannot be applied strictly, since cutting edges are often curved and tool-normal machining process means that chip flow can occur along many directions on the rake face during DTM processing. Again, experimental data (of diamond on diamond) is available for some guidance [33]. It is not clear whether such data for rubbing of various engineering materials on diamond, under machining-related tribological conditions, is readily available.

Besides the nose radius geometry, another important consideration is the cutting edge radius. This portion of the cutting tool is the weakest and is subject to micro-fracture and cleavage leading to wear and edge radius enlargement. Investigations remain to be carried out to study edge geometries and on the best way (e.g. crystallographic planes, directions, geometries

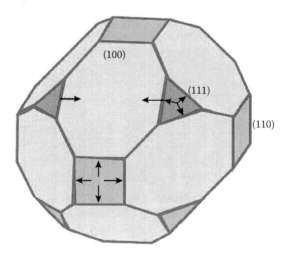

FIGURE 4.5
Schematic summarising the various soft crystallographic directions (shown with black arrows) to abrade diamond.

such as chamfer, land, etc.) to shore up the edge strength against chipping. Fabrication challenges need to be addressed to incorporate such fine edge control in diamond crystals.

The other consideration in the diamond crystal is the surface finish that is imparted on the various tool surfaces. The crystallographic dependence on resistance to abrasive wear is also responsible for varied roughness patterns on the diamond surface. Ideally, the diamond tool rake face should be very smooth (sub-nanometric roughness) and be isotropic – this way any chip flow direction can be supported. However, attaining this requires several steps of polishing processes.

Sub-surface damage imparted to the crystal by the abrasive fabrication process is also an important consideration in the performance of the cutting tool in the DTM process. Sub-surface damage can be detected by etching the diamond with suitable chemicals with thermal assist in an atmosphere rich in oxygen. Each abrasive process leaves behind its signature in the form of surface lay patterns and also sub-surface damage and stresses. Polishing should be undertaken in various steps so that the damage induced in each step is removed by the subsequent processing step.

4.4 Tool Geometry

Diamond tools are commercially available in a variety of macro-geometrical shapes – both standard shapes and customised shapes. Some common geometries routinely used are shown in Figure 4.6.

The most common cutting edge shape is triangular with the nose radius at one apex of the triangle. Other edge shapes include:

- Flat rectangular shape used in grooving applications with customised groove widths down to a few micrometres
- Two straight cutting edges meeting at an obtuse angle – used for nano-milling

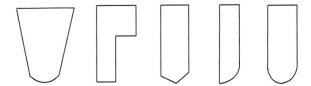

FIGURE 4.6
Schematic showing (top view of the rake face, showing cutting edges) various geometries in which diamond tools are available.

- Single arc cutting edges – used for milling
- Multiple discontinuous connected arcs – used for nano-milling applications
- Elliptical arc shape

Besides the overall cutting edge profile, other geometrical parametres of interest are: Top Rake Angle, Cylindrical/Conical Edge, Tool Nose Radius, Cutting Arc, Offset Angle, Included Angle, Front Clearance, Second Clearance, Primary Depth, Diamond Depth, Total Cutting Height and Tool Nose Waviness. Clearance angles are very important especially in chiseling of complex micro-geometries using fast tool servo (FTS)/slow tool servo (STS) applications, as they limit the degree of interaction of the cutting tool with the previous machined surface. Having large clearance angles is beneficial to making complex-shaped micro-arrays in optics, but weakens the cutting edge significantly. Having a strong cutting edge is especially important, since diamond is brittle and interrupted cutting causes significant stresses at the edge. It is common to use zero rake angles with SCD tools. However, often negative rake angles are needed for machining brittle materials such as silicon and germanium to maintain ductile regime machining. The negative rake can sometimes be obtained in the diamond crystal itself; if this is not possible, the tool holder seating surface is inclined to provide the needed angle.

The cutting edge should be extremely smooth and its profile should not deviate significantly with edge profile tolerances often in fractions of micrometres. This is referred to as controlled waviness of the tool (Figure 4.7).

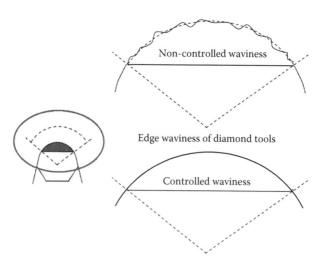

FIGURE 4.7
Waviness in the tool nose radius.

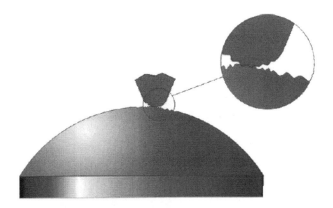

FIGURE 4.8
Tool form error gets transferred to the work machined surface.

Commercial suppliers also supply tools without the waviness being controlled for less demanding applications. Any form errors on the cutting tool will duplicate on the machined surface (Figure 4.8).

4.5 Diamond Tool Fabrication

The typical structure of a SCD tool used in DTM is shown in Figure 4.9. The tool consists of a single crystal of shaped diamond that is brazed onto a tool holder. Such a structure is commonly seen in face-turning, grooving and fly-cutting (ultra-precision milling).

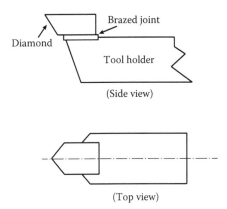

FIGURE 4.9
Schematic of the typical structure of a single crystal diamond tool and holder.

Aspects related to the fabrication of the diamond crystal itself are discussed in an earlier section in this chapter. The key steps are: determining the type and quality of the diamond, its crystallographic plane and directions to chisel out the tool, holding the crystal in appropriate orientations and shaping the crystal surfaces to provide the desired geometry for the rake face, nose radius and edge radii, while imparting prescribed surface finish to the rake face, waviness to the cutting edge and uniformity in edge radius.

Determining the crystal orientations is a key step, once the diamond crystal is chosen. Synthetic diamonds are often available with a flat face of known crystal orientation. However, natural diamonds have to be subjected to single crystal x-ray diffraction measurements to determine crystal orientations. Once the needed planes are identified in the x-ray machine, the crystal needs to be positioned at that orientation in a fixture ready to be further processed – this involves a suitable fixture design to perform this action. Once the crystal is positioned in the fixture, a diamond abrasive grinding wheel can be used to grind the top crystal surface to create the flat rake face. The clearance faces are then ground using a bonded abrasive (diamond) wheel. Next delicate task is to create the nose radius using a rotating fixture. Controlling the edge waviness and further polishing steps (e.g. lapping) remain company trade secrets that have been developed only by experimental trial and error with little published records.

The tool holder that houses the diamond crystal is made of steel or a type of molybdenum alloy. The alignment of the nose radius center and the center of the tool holder is important; also critical is the orientation of the rake face plane with respect to the tool holder base. The tool holder is often ground and polished to achieve needed flatness and perpendicularity, and sometimes also coated with Ni/Cr.

Brazing the diamond onto the substrate is a key step in the tool fabrication process. The brazed joint strength is obviously important for the robust performance of the tool during DTM machining. Various filler materials for the brazed joint are now available, with brazing temperatures exceeding 1000°C. At such temperatures, diamond can graphitise and lose its desired structure. One way to overcome this problem is to undertake brazing at lower pressures (vacuum); lowering the pressures and introducing inert gases increase the graphitisation temperature. For example, graphitisation temperatures increase from 1000°C to as much as 1600°C when pressures are lowered and inert atmosphere is introduced. However, some local graphitisation at the brazed interface may be desirable, since it ensures good joint strength. Metals do not diffuse and are not dissolved in the diamond. However, diamond does form carbides relatively easily. Researchers have discovered that by adding Ti, Cr, Ta and/or Si to brazing alloys, stable carbides are formed which wet and allow the brazing of diamond to many materials. These materials are known as active brazing filler metals. Alloy systems including Cu-Ag-Ti, Cu-Sn-Ti or Ni-Cr-B are common brazing filler metals for joining diamonds to metals and ceramics.

4.6 Tool Wear

Just as in conventional machining, diamond tools wear out during the DTM process and become unusable. This finite life of the tool dictates significantly the economics of the DTM process; tooling costs are enormous barriers in successfully processing several work materials such as silicon, germanium etc. Various types of wear in conventional machining include flank wear, crater wear, edge chipping etc. The differences between tool wear in conventional machining and in DTM process are the degree of wear and type of wear. The DTM process is highly sensitive to wear – hence, only small amounts of tool wear can be tolerated, in comparison to conventional machining. Crater wear is seldom seen in diamond tool wear, while edge radiusing, edge chipping and groove formation in the flank face are more common. High temperatures and chemical diffusion of carbon atoms cause rapid wear of the diamond tool when used to machine ferrous alloys. Various techniques were attempted to reduce such wear so that ferrous alloys can be processed using diamond tools and the DTM process. Commercial success, however, is still elusive barring a few successful applications. Tool wear in the DTM process not only directly influences the condition of the machined surface but also indirectly by affecting the mechanism of cut. Ductile regime material removal conditions, for example, can be very sensitive to tool wear.

Tool wear, while observed using scanning electron microscopy, is often measured using optical metrology techniques. One such technique used and reported is coherence correlation interferometer (CCI) (e.g. Taylor Hobson CCI 6600). This method involves measuring variations in interference fringes as the vertical position of the optical elements above the surface being scanned is changed. Sample images of new and worn tools measured using this technique are shown in Figure 4.10. Waviness is determined by measuring the nose radius profile (Figure 4.11) and its deviation from an ideal circular arc.

Such CCI measurements can be made during the life of the diamond cutting tool to observe how the tool wear develops with machining time (Figure 4.12). In this particular example, the cutting tool was used to machine single crystal silicon and the waviness progressed from about 3 μm in the unused condition to over 12 μm at the end of 200 cycles (276 minutes of machining time). The wear in this case is predominantly of the tool edge chipping, with the amount of chipping gradually increasing in depth into the rake face and widening along the cutting edge. Figure 4.13 shows a typical correlation

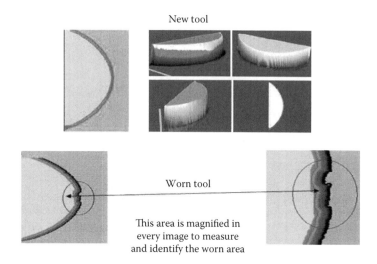

FIGURE 4.10
Coherence correlation interferometry images of a diamond tool edge.

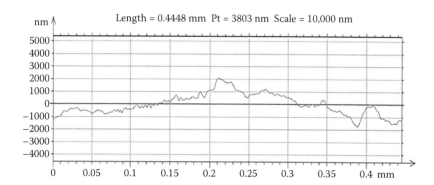

FIGURE 4.11
Profile of the nose radius used to determine waviness.

between the change in the waviness of the tool cutting edge due to wear and machining time.

Tool edge radius, often in tens of nanometres in magnitude or less, and its increase in magnitude with machining time are more challenging to measure. Rough estimates have been obtained using side images of a straight cutting edge from scanning electron microscopy; these give only the imaged plane measurement. More sophisticated techniques such as atomic force microscopy (AFM) have been attempted to measure the edge radius, both directly and indirectly (Figure 4.14). It is challenging to locate the edge during AFM measurement and AFM tip interactions with the edge can cause measurement errors.

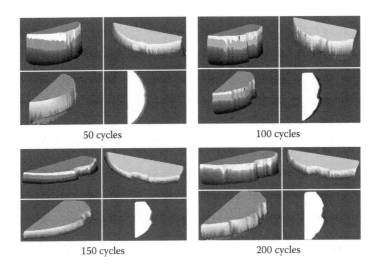

50 cycles   100 cycles

150 cycles   200 cycles

FIGURE 4.12
Diamond tool wear development.

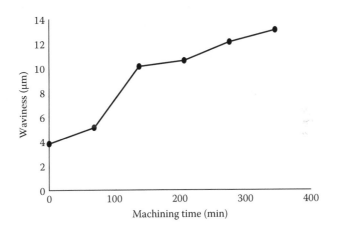

FIGURE 4.13
Progress of waviness with machining time.

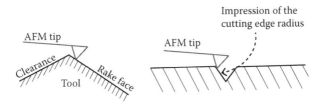

FIGURE 4.14
AFM measurements of tool edge radius.

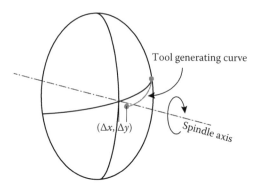

FIGURE 4.15
Tool setting error can lead to errors in the surface generated.

4.7 Tool Setting in DTM

Errors in setting the diamond tool on the DTM equipment can cause errors in the surfaces produced. The classic case is the so-called ogive error (Figure 4.15). In the classic DTM face-turning process scenario, the work-piece revolves around the spindle axis, and the path of the tool on one side of the axis generates an axially symmetric surface. The ogive error arises when one end of the tool path motion does not coincide with the central rotational axis. This occurs especially when it is difficult to determine accurately the height of the tool tip and the centre of the tool nose radius.

4.8 Summary

The single crystal diamond has been the flagship cutting tool for the DTM process. Despite its high hardness, it has provided the flexibility to carve out rake face and cutting edges on selected crystallographic planes and orientations. It has also provided the ultra-sharp cutting edge needed for successful processing of several engineering materials with the desired optical quality surface. While the solution to machine ferrous alloys has still been challenging, diamond has addressed most of the optical needs. No reported progress has been made with growth of large single crystals of other materials such as cubic boron nitride for use as cutting tools to process materials that are not DTM-machinable with diamond.

4.9 Unsolved Problems

1. Describe the various steps involved in the diamond tool fabrication process.

2. Conceive the design of a fixture which allows transfer, without loss of orientation, of the diamond crystal from the single crystal x-ray diffraction equipment to the abrasive polishing machine.

3. Draw a thumbnail sketch of a diamond tool and label some important geometrical parametres.

4. Describe the various soft and hard crystallographic directions of abrasion in the diamond cubic structure.

5

DTM Process Parametres and Optimisation

5.1 Introduction

Diamond turn machining possesses the capability to generate surfaces with optical quality and with nano-regime control on size, shape and surface finish. Components for optical mold, astronomical telescope, IR optics, optics for avionics head gear and laser optics use materials like nickel, aluminium alloy, silicon, germanium, polymers and copper. A wide range of such materials for varied applications can be machined extensively by diamond turn machining. Machining of these materials on diamond turn machines requires single crystal diamond tools of specific grades and tool geometries. Appropriate process parametres like speed, feed and depth of cut need to be employed to get the desired defect-free surfaces. Twin requirements of generating defect free surfaces with controlled size, shape and surface finish as well as minimising of tool wear pose severe problems during diamond turn machining.

To tackle such problems, it is important to have a thorough understanding of the role of various input parametres affecting the diamond turn machining process and their relevance to different output parametres [34]. In this direction, a brief discussion is carried out in this chapter.

5.2 Diamond Turn Machining Process and Parametres

Every diamond turn machine has its own characteristic behaviour and doesn't perform similarly for a given tool–work material combination. For example, the natural frequency of the machine which causes chattering varies from machine to machine. Therefore, it is important to map the behaviour of a diamond turn machine for various process conditions and optimise them. Figure 5.1 shows the scheme of a diamond turn machining process with various inputs and

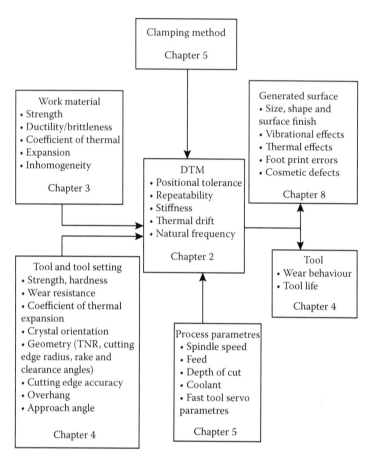

FIGURE 5.1
Diamond turn machining scheme.

output parametres. During the diamond turn machining process, the following sequence is generally adopted:

- Selection of a diamond turn machine whose characteristics like positional tolerance, stiffness, thermal drift etc. are known *(Known behaviour)*;
- Material on which the surface is to be generated *(Known properties)*;
- Selection of tool grade, geometry and the crystal orientation suitable for the component geometry and material *(Known parametres)*;
- Clamping method for the work-piece considering elimination of footprint error and stability of holding *(Known methods)*;

- Selection of optimised process parametres like speed, feed, depth of cut, cutting direction, coolant, parametres of fast tool servo etc. *(Optimised values).*

Following the above steps leads to achieve the below-mentioned outcomes during diamond turn machining:

- Generate surfaces free from cosmetic defects, vibration and thermal effects; and control their size, shape and surface finish value. (Process should be controlled to achieve this objective)
- Predictable and minimum tool wear. (Wear prediction models are essential)

With these objectives in mind, various process parametres and their effects will be discussed in the following sections.

5.2.1 Spindle Speed

In diamond turn machining, the spindle holds either the work-piece or the fly cutter. During the machining process, constant spindle speed is employed in order to prevent acceleration of the spindle and to avoid the effect of inertial force on the machined surface. Therefore, the rotational speed of the spindle is kept constant. This results in the cutting velocity changing from zero at the axis of the spindle to maximum at the largest radial distance from the axis of the work-piece. The spindle speed facilitates the following two outcomes:

- Enables removal of material by providing relative movement between the work-piece and the tool. Thereby, it helps to achieve the desired shape, size and surface finish of the component;
- Enables maintaining the required level of productivity by controlling the material removal rate.

Unlike other similar machining processes, where higher levels of speed ensure better surface quality, diamond turn machining does not demand higher spindle speed owing to the use of a very sharp cutting edge. A typical example is machining by fast tool servo, which is carried out with a spindle speed of a few revolutions per minute. However, to achieve a higher material removal rate, a higher spindle speed is necessary. Increasing the spindle speed beyond a certain value on the contrary increases the tool wear and thereby causes more size and shape variation on the work-piece; tool wear also blunts the cutting edge and changes the material removal mechanism. Many researchers published the correlation between the spindle speed and the achieved surface finish value for different materials [9,35–39]. In the reported instances, the variations on surface finish value with changing speed are not only due to the change in cutting mechanism arising out of tool wear alone, but also by a number of other factors like induced vibrations from the work material and from the machine elements, increasing thrust force with tool wear, etc. Hence, optimum spindle speed is to be arrived at after carrying out trial machining on given material. Figure 5.2 shows a typical relationship between the spindle speed and surface finish. Table 5.1 shows the spindle speed values for diamond turning of some engineering materials.

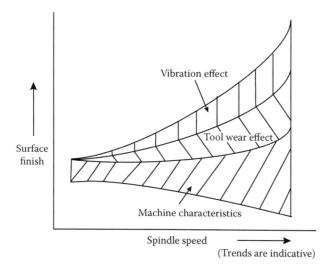

FIGURE 5.2
Effect of spindle speed on surface finish.

TABLE 5.1

Optimum Spindle Speed for Various Materials

Material	Diameter of Work-Piece in mm	Top Rake Angle in Degrees	Tool Nose Radius, mm	Spindle Speed, rpm
Copper	50	0	0.5	2000
Aluminium alloy	300	0	0.5	1500
Nickel	15	0	0.5	1000
Silicon	150	−25	6	1000
Germanium	50	−25	0.6	1000
PMMA	50	0	0.1	2500
Brass	10	0^0	0.5	2500

5.2.2 Feed Rate

The primary objective of feed in diamond turn machining is to shape the component in the desired fashion to achieve size, shape, surface finish, reflectivity, etc. by controlling the feed path. During the course of machining, lay pattern in the form of micro-helical grooves is generated on the surface and it impairs the surface quality. In general, feed rate affects the following factors:

- Cutting mechanism
- Unit removal (UR) of material
- Magnitude of cutting force
- Tool wear
- Quality of the machined surface in terms of surface finish, vibration effects, etc.
- Cycle time

The unit removal of material, which is proportional to the cross-sectional area of the chip, changes with feed. Accordingly, the material removal rate increases and cycle time decreases; optimisation of feed for cycle time becomes important when machining a large sized component as well as a large batch of components.

One of the most important requirements of diamond turn machining is to deliver optical quality on the machined surfaces. To achieve this objective, it is necessary to control various process parametres including feed rate. It is well known that the surface finish value is given by the following equation in any single point machining process:

$$\text{Surface finish (PV)} = f^2/8R \tag{5.1}$$

where f = feed per revolution and R = tool nose radius.

However, this equation has been modified by many researchers and some of these results are shown in Table 5.2.

Feed force generated during machining increases with feed rate and is given as:

$$\text{Feed force } F_f = C_f\, A_f\, V_f = C_f\, d\, b\, V_f = C_f\, d\, b\, n\, f \tag{5.2}$$

where C_f = constant; A_f = cross-sectional area of chip; d = depth of cut; b = width of chip; V_f = feed velocity; n = rotational speed; and f = feed rate.

TABLE 5.2

Achievable Surface Finish Values in DTM

Surface Finish Value	References
Peak to valley, $R_{max} = (f^2/8R)$ where f = feed rate, R = tool nose radius	[9]
$R_{th} = (f^2/8R) + t/2\,(1 + t\,R/2)$ where t = min. undeformed chip thickness Theoretical average roughness, $R_{th} = (f^2/8R)/4$	[35] [36]
$R_{th} = (f^2/8R) + t/2\,(1 + t\,R/f^2)$ $R = A + a_2\,(f^2/8R)$ where $A = a_0 + a_1/f$; $a_0 = 12.4109$; $a_1 = 4.0529$; $a_2 = 0.2317$	[37] [38]
$R_{th} = (f^2/8R) + t/2\,(1 + t\,R/2) + k_1\,k_2\,r_n\,H/E$ k_1 = coefficient in relation to the elastic recovery, k_2 = coefficient denoting the size effect, H = Vicker's hardness, E = modulus of elasticity, r_n = tool cutting edge radius	[39]

Tool wear increases with the feed force and results in increased size and shape errors on the generated surface. Figure 5.3 schematically shows the effect of feed on the above-mentioned outcomes. Figure 5.4 shows the effect of depth of cut and feed on the length of the nano-cutting region as indicated by Region 1. In general, increasing length of this region deteriorates the surface finish value.

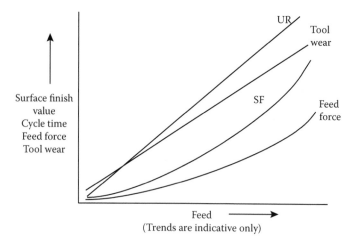

FIGURE 5.3
Effect of feed on various output parametres.

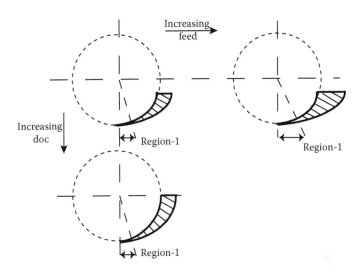

FIGURE 5.4
Effect of feed and depth of cut on nano-cutting length.

5.2.3 Depth of Cut

Unlike feed, influence of depth of cut (DOC) on the surface quality is minimal. Referring to Figure 5.4, when the depth of cut is changed, the proportion of the chip length corresponding to the nano-cutting region (Region 1) remains more or less the same and hence the surface finish value remains unaffected. However, the material removal rate increases with DOC.

5.2.4 Tool Shank Overhang

Change in the length of overhang of the tool shank changes its stiffness and affects the surface quality of the surface generated [40]. Figure 5.5 shows the relationship between achievable surface finish and tool shank overhang. Until the tool shank overhang reaches an optimum length, its stiffness is more than the loop stiffness value and after crossing the optimum length, the stiffness decreases; hence, deterioration of the surface finish below and above the optimum tool shank length is noticed. Single crystal diamond tools are brittle and fragile. While machining, variation in the tool shank overhang from its optimum value leads to a damaged cutting edge. More often, the tool tip breaks.

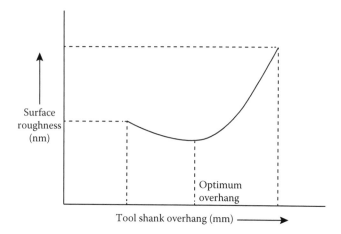

FIGURE 5.5
Effect of tool shank overhang on surface finish.

5.2.5 Coolant

Heat generated at the machining zone is carried away by

- Chip
- Tool
- Work-piece
- Coolant

As the cross-sectional area of the chip is much smaller, larger proportion of the heat generated is carried away by it. Similarly, the diamond tool being an excellent heat conductor, a substantial amount of heat is transferred through it. As the heat transferred to the work-piece is likely to cause thermal damage to the machined surface, efforts should be made to transmit minimal amount of heat into it and a larger proportion of the heat should be transferred through other sources. Heat transferred through the chip and the tool is limited by the volumes of work and tool materials and by their respective thermal conductivities. Therefore, removing the maximum amount of heat from the cutting zone by coolant is the most effective way of preventing damage to the machined surface. In order to achieve this objective, in most of the applications mist coolant is employed, which can carry away a large amount of heat as latent heat. The chips generated are light in weight, have longer length and often are in powder form. Hence, they have more affinity to stick to the finished surface and may damage it due to:

- Adhesion, where part of it gets cold welded to the generated surface and it becomes difficult to remove these chips, without impairing the quality of the surface (Figure 5.6a);

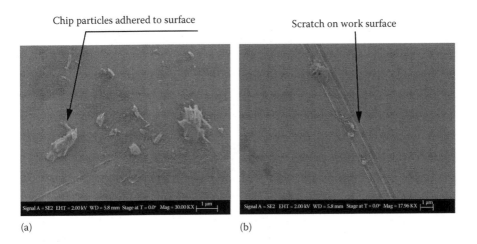

FIGURE 5.6
(a) Adhesion of chips on the finished surface. (b) Digs and scratches on the finished surface caused by chips.

- Abrasion, which generates digs and scratches on the machined surface and causes cosmetic defects (Figure 5.6b);
- Interference, where long chips get curled on the tool and break the tool tip.

To overcome such problems, the orientation of the coolant nozzle is maintained to blow away the chips from the machined surface and coolant is mixed with suitable additives to minimise the affinity between the chip and the machined surface.

5.2.6 Clamping Method and Footprint Error

The clamping force induces undesirable strain on the work-piece and on its removal, the strain is released and the machined surface gets distorted. The elastic deformation on the machined surface on release of clamping is termed the *footprint error*. Generally, the two methods of clamping that are practiced in the diamond turn machining industry are:

- Flexible clamping, either by direct mounting of the work-piece on the vacuum chuck or using a suitable fixture;
- Rigid clamping of the work-piece with a suitable fixture

In the first method, the work-piece is mounted on the machine spindle using vacuum chuck. The work-piece is either directly mounted on the vacuum chuck or the fixture holding the work-piece is mounted on the vacuum

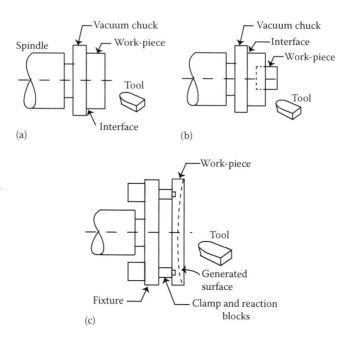

FIGURE 5.7
Clamping methods. (a) Direct vacuum clamping of work-piece. (b) Vacuum clamping of the fixture with work-piece. (c) Rigid clamping method.

chuck. Figure 5.7 shows both these arrangements as well as the rigid clamping method. In the case of vacuum chuck clamping, the work-piece/fixture has the flexibility to accommodate the variation in the cutting forces arising due to the material inhomogeneity or some other reason. As a result of the micro slips taking place at the interface of vacuum chuck and work-piece/fixture, the effect of shock loading on the cutting edge is minimised and the tool life is enhanced. In the case of rigid clamping, the work-piece is held rigidly using fixtures without having any flexibility. Whenever the component size is large or non axi-symmetric, rigid clamping becomes the only option.

5.3 Vibration Related Issues

As discussed in Chapter 2, vibration plays a destructive role while generating surfaces in diamond turn machining. Both material-induced vibrations and machine-induced vibrations are transferred to the interface of tool and work-piece and leave their signatures on the surface generated [14]. Sources of the vibration include vibration arising from the spindle bearing element, imbalance

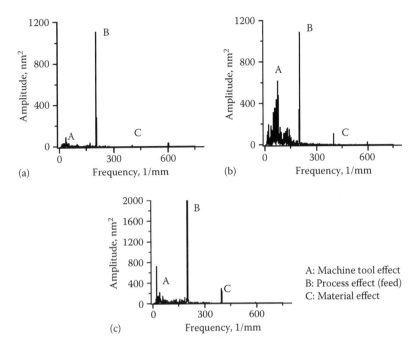

FIGURE 5.8
PSD from surface finish profile of diamond turned surface. (a) Process dominated. (b) Machine tool effect and process dominated. (c) Machine tool, process and tool wear dominated.

in the spindle rotation, fluctuations in pneumatic and electrical power supply to the spindle drive, table bearing elements, material inhomogeneity, cutting force and external vibrations. Online analysis of the vibrational signals by power spectral density (PSD) gives a fair idea about the sources of the vibration. Figure 5.8 shows a typical PSD of the surface finish profile obtained from a diamond turned surface. Initially, effects of the machine and the process are found to be dominating factors for generating vibration and subsequently the tool wear is found to be dominating. Stability lobe diagrams drawn with the data of tap testing with a delicate hammer for frequency response are some of the most useful tools to determine various stable speeds for a given arrangement of machine, cutting tool, tool holder and tool overhang.

5.4 Thermal Issues in Diamond Turn Machining

During the machining process, input power is converted into heat energy and is dissipated through chip, tool, coolant and work-piece. The transmitted heat has a damaging effect on the tool as well as on the work-piece [41]. Tool wear gets accelerated and subsequently affects the quality of the generated surface due to pressure and heat at the machining zone. If the heat transmitted to the work-piece is not removed quickly, it damages the surface integrity of the machined surface [42]. Types of damages include:

- Change in the mechanical and metallurgical properties of the surface layer including:
  - Hardness
  - Residual stresses
  - Optical property
- Swelling of material affecting the dimensional tolerance and shape error
- Surface finish
- Colour

The intense heat generated while machining moves along with the point of cutting. The moving heat source can be modelled and its effect can be visualised for any material which can become a guiding tool in optimisation. The effect of the generated heat is more severe in the case of materials having poor thermal conductivity. For example, a thermal effect is more severe on polymer materials than on materials like copper, aluminium alloy, etc. Materials with low thermal diffusivities (like germanium, which is extensively used for thermal imaging camera) are very sensitive to the thermal effects and need extensive preventive measures to avoid thermal damage to the finished optical surface. Preventive measures include:

- Selecting proper machining sequence for enabling the finishing pass to remove the previously thermally damaged layer;
- Preventing generation of excessive heat due to tool rubbing with the finished surface

5.5 Optimisation of DTM Parametres

A number of factors affect the quality of the diamond turned surface. As these factors dynamically affect the tool wear, optimisation of the process parametres to achieve the desired surface quality as well as to enhance the useful tool life is essential. Simultaneously, vibration and thermal effects are to be accounted for, while achieving the desired surface quality.

Conducting extensive experimentation for optimisation is expensive for smaller batch production; hence, guidelines and data available from the literature are to be used for optimisation. Table 5.3 shows the factors that need to be considered when optimisation is carried out.

TABLE 5.3

Factors Requiring Optimisation

Factors	Consideration/Technique
Work material	Fixed by designer
Machine	Capacity and capability of the machine
Tool grade and geometry	Tool manufacturer catalogue and data base
Tool overhang	Stiffness, vibration
Method of clamping	Footprint error
Speed	Stability lobe for the system and tool life
Feed	Surface finish and tool life
Depth of cut	Tool life and productivity
Coolant and nozzle position	Surface quality and tool life

5.6 Summary

This chapter discusses the effects of various diamond turn machining process parametres on different outcomes, viz: surface quality, tool life etc., and various considerations while optimising the parametres and the effects of vibration and thermal and clamping methods on the machined surface.

5.7 Sample Solved Problems

Example 1. Find the locus of the diamond tool and the equation of surface generated during diamond turn machining of a convex-hemispherical shaped surface on a copper shaft. Diameter of copper shaft = 15 mm, nose radius of the diamond tool = 1.5 mm and cutting arc angle = 120°. Neglect the effect of the machine, tool stiffness and elastic recovery of work material. Assume that the Z-axis is the axis of rotation and the X-axis is the radial direction. What would be the initial and last coordinates of the diamond tool for circulation interpolation during this process?

Solution 1: Refer to Figure 5.9.

Radius, $R = 15/2 = 7.5$ mm; radius of the diamond tool, $r = 1.5$ mm

AB is a circular arc (quarter size of a circle) with center O and radius R; hence, the equation for path AB can be written as:

$$X^2 + Z^2 = R^2 \tag{5.3}$$

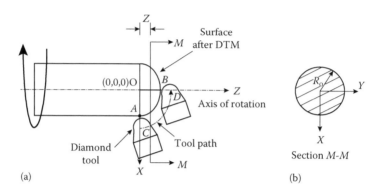

FIGURE 5.9
(a) Schematic for tool movement along a circular interpolation to machine hemispherical shape and (b) cross-sectional view along *M-M* on the *X-Y* plane at a specific *Z*-value.

Similarly, the equation of path *CD* can be written as;

$$X^2 + Z^2 = (R+r)^2 \text{; or } X^2 + Z^2 = (7.5+1.5)^2 = 81, \text{ or } X^2 + Z^2 = 81 \quad (5.4)$$

This is the locus of the diamond tool.

The initial point of the tool $C = (9,0,0)$ and the final point of the tool $D = (0,0,9)$. R_o can be written as the value of X at the specified value of Z using Equation 5.3:

$$R_o = (R^2 - Z^2)^{0.5} \quad (5.5)$$

Using Equation 5.5 the circle of Figure 5.9b can be written as:

$$X^2 + Y^2 = (R_o)^2 = (R^2 - Z^2) => X^2 + Y^2 + Z^2 = R^2 \text{; or } X^2 + Y^2 + Z^2 = 56.25$$

This is the equation of the DTM turned surface.

Example 2. Find the equation of the diamond turned surface, which is generated after diamond turn machining of a convex-hemispherical shaped surface on a copper shaft, if the diamond tool is misplaced radially inside by 0.010 mm. The diameter of copper shaft = 15 mm, nose radius of the diamond tool = 1.5 mm and cutting arc angle = 120°. Neglect effect of machine, tool stiffness and elastic recovery of work material. Assume that the Z-axis is the axis of rotation and the X-axis is the radial direction.

Solution 2: Refer to Figure 5.10.

From Figure 5.10a, the equation of the circular path *AB* can be written as:

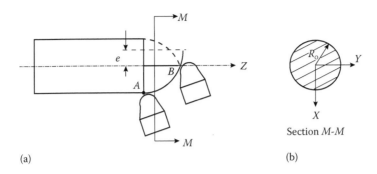

FIGURE 5.10
(a) Schematic for tool movement along a circular interpolation to machine hemispherical shape after introducing eccentric error on the tool and (b) cross-sectional view along M-M on the X-Y plane at a specific Z-value.

$$(X+e)^2 + Z^2 = R^2 => (X+e)^2 = R^2 - Z^2 => (X+e) = (R^2 - Z^2)^{0.5} => X$$
$$= (R^2 - Z^2)^{0.5} - e \quad (5.6)$$

This X can be represented as the radius of Figure 5.10b. Hence, the equation of the diamond turned surface can be formulated using Equation 5.6 as

$$X^2 + Y^2 = (R_o)^2 = \{(R^2 - Z^2)^{0.5} - e\}^2 = (R^2 - Z^2) + e^2 - 2e(R^2 - Z^2)^{0.5} \quad (5.7)$$

5.8 Questions and Problems

Q1: Find the locus of the diamond tool and the equation of the surface generated during diamond turn machining of a convex-hemispherical shaped surface on a copper shaft. Diameter of copper shaft = 20 mm, nose radius of the diamond tool = 1.0 mm and cutting arc angle = 120°. Neglect the effect of machine, tool stiffness and elastic recovery of the work material. Assume that the Z-axis is the axis of rotation and the X-axis is the radial direction.

Q2: Find the locus of the diamond tool to generate a concave mirror on aluminium with 300 mm of 'radius of curvature'. Diameter of aluminum mirror = 15 mm, nose radius of the diamond tool = 2.0 mm and cutting arc angle = 120°. Neglect the effect of machine, tool stiffness and elastic recovery of the work material. Assume that the Z-axis is the axis of rotation and the X-axis is the radial direction. What would be the initial and last coordinates?

Q3: Find the equation of the diamond turned surface that is generated after diamond turn machining of a convex-hemispherical shaped surface on a copper shaft if the diamond tool is misplaced radially inside by 0.050 mm. Diameter of copper shaft = 15 mm, nose radius of the diamond tool = 1.5 mm and cutting arc angle = 120°. Neglect effect of machine, tool stiffness and elastic recovery of the work material. Assume that the Z-axis is the axis of rotation and the X-axis is the radial direction.

6

Tool Path Strategies in Surface Generation

6.1 Introduction

For the DTM process to generate a new surface, a cutting motion along with feed motion is required as in any machining process. The cutting motion provides the high momentum interference between the tool and work-piece to shear or fracture and remove material, while the feed motion continually provides new material to cut. The cutting motion is normally provided in the form of the work-piece mounted on a rotating spindle, while the feed is provided to the cutting tool at a much lower speed movement in an orthogonal direction. These two motions when coordinated in different ways provide various geometrical surfaces in the DTM process.

In a general sense, surfaces generated by diamond turn machining can be classified into rotationally symmetric (Figure 6.1a) and rotationally asymmetric ones (Figure 6.1b). Within asymmetric ones, there can be surfaces that have very fine micro-textured features (Figure 6.1c), such as those in diffractive optic elements, for example. The symmetry axis is the axis around which the part is normally rotated in the spindle during the DTM process. The surface shown in Figure 6.1b may seem symmetric at first glance, but is not rotationally symmetric around the axis; if one rotates the surface around the central axis, the surface features are seen to change with angular position at many radial positions.

Rotationally symmetric parts are commonly produced by conventional turning processes. DTM can also manufacture such rotationally symmetric parts. Additionally, nowadays DTM is routinely used to produce rotationally asymmetric parts. This requires a different set of coordinated motions of the tool and the work surface. While producing rotationally symmetric parts, the tool position can be unrelated to the rotational angular position of the part, whereas while producing rotationally asymmetric parts, the tool position is tied to the angular rotational position of the part in the rotating spindle. The next two sections explain the tool paths for these two types of shapes made in DTM.

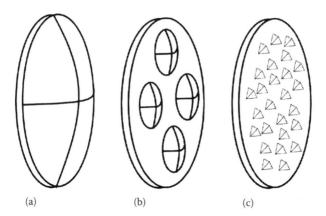

FIGURE 6.1
Example of (a) a rotationally symmetric shape, in contrast to (b) a rotationally asymmetric surface – the otherwise planar surface has several convex lens protrusions placed off-centered. (c) A part surface filled with tiny features that are tens of micrometers in size.

6.2 Tool Paths for Symmetric Macro Shapes

The diamond turn machining process normally involves the face turning operation just as in any typical turning process. In a normal turning process, the work surface spins about the spindle axis with the tool feed motion in a direction perpendicular to this axis. The tool feed motion path typically lies in a plane (XZ) containing the spindle axis (Z) (Figure 6.2).

Thus, a typical DTM is a 2-axis simultaneously controllable machine. The tool feed motion is usually not synchronised with the spindle rotation; the spindle either runs at a constant rotational speed or in a monotonically

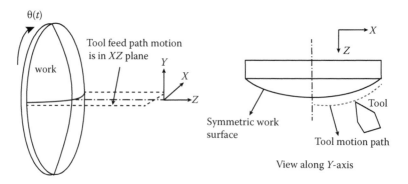

FIGURE 6.2
Tool motion path for symmetric shapes involves spindle rotation to provide cutting motion while the tool moves in the XZ plane to generate the necessary geometry.

increasing or decreasing speed to maintain constant tangential speed at the cutting tool tip. Because of this lack of synchronicity, machining features produced naturally tend to be rotationally symmetric (Figure 6.1) about the axis. This is the normal arrangement in a simple DTM setup, which also leads to production of shapes symmetrical about the axis. Such shapes can be spherical or aspherical (elliptic, parabolic, etc.), but symmetric about the axis.

Let us now consider on how material is removed to create the desired symmetrical surface. Consider first creating a flat planar face. The face turning operation removes an entire disc-shaped volume of material from the front face of the work-piece in the form of a slowly peeling spiral of material (Figure 6.3). The spiral can be better visualised in the following way: If you can imagine the spindle to be brought to a stand-still, but allow the tool to have both rotational cutting and feed motion, then you may see that the tool traverses on the face of the work material in the form of a spiral, starting from the outside and slowly converging toward the center (Figure 6.3). This material removal in a spiral fashion holds true even when the face is not flat but has a curvature (e.g. a sphere) – like spirally peeling the skin of an orange. This path can be mathematically quantified as follows:

Consider the DTM operation of facing a flat surface such as in Figure 6.3. Assuming that the spindle revolving speed is N (rev/s), and the radial

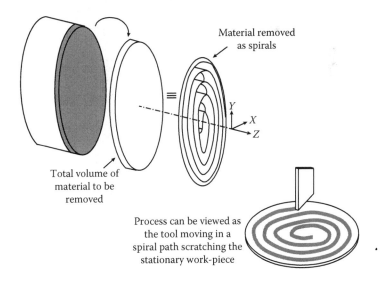

FIGURE 6.3
Schematic illustrating how face turning operation, the basic process in DTM, removes material in the form of a uniform thickness spiral to produce symmetrical shapes – in this case a planar surface. This form of material removal applies even when the face is not flat but has a curvature (e.g. spherical surface).

feed of the tool is f mm/rev, the spiral path of the cutting tool would be an Archimedes spiral of the form (in polar notation):

$$r = r_{ref} - \frac{f}{2\pi}\theta; \tag{6.1}$$

where r_{ref} denotes a faraway position that the tool moves radially inwards from. Hence, the x and y coordinates of the spiral path can be found as (origin at the center on the spindle axis at a suitable point):

$$x = r\cos\theta = \left(r_{ref} - \frac{f}{2\pi}\theta \right)\cos\theta \tag{6.2}$$

$$y = r\sin\theta = \left(r_{ref} - \frac{f}{2\pi}\theta \right)\sin\theta \tag{6.3}$$

assuming that the tool starts from an outer point (defined using R_{ref}) and spirals inwards. For the flat surface, z is zero or a constant.

For a spherical surface (of radius R) of the type shown in Figure 6.2, the path will be a spiral in three dimensions, with an additional varying z-coordinate given by (assuming that tool starts from a reference point, z_{ref})

$$z = z_{ref} \pm \sqrt{R^2 - r^2} \tag{6.4}$$

It is clear that since the feature is rotationally asymmetrical, z is not a function of θ. For a parabolic surface of revolution characterized by constant a, the following would be the z-coordinate variation with r, again independent of θ:

$$z = z_{ref} \pm \frac{r^2}{4a} \tag{6.5}$$

For practical path execution with a rotating spindle, and radial feed path, set $\theta = 0$. Then $r = x$, and the controller determines the position x and then based on this, the z-coordinate for tool motion.

For non-analytical surfaces, the following approach can be undertaken. On the projected flat surface (XY plane), based on a feed rate in the x-direction, an Archimedes spiral path is adopted and a discrete set of points, (r, θ) in polar coordinates, identified. At these points, the z-coordinate is determined based on the given data or non-parametric relationship. Sufficiently close points (r, z) (θ is set to zero since the spindle is rotating) can then be used along with linear interpolation (or other interpolations, such as a spline-based, based on controller sophistication and encoder resolution) to fill in the gaps between the points for tool movement.

This spiral motion path concept is important to understand, since this in turn will make it easier to visualise how, in the discussion that follows in the next section, asymmetrical shapes are created by synchronising tool feed motion (in the Z-axis) with the spindle rotation (C-axis). Note that only finishing path motions are described here. Rough cut motion paths and associated algorithms are techniques that can be adopted from standard CNC literature.

6.3 Tool Paths for Producing Asymmetric Macro Shapes

Consider the asymmetrical shape shown in Figure 6.1b. What kind of tool feed motion paths are needed to produce this shape while providing cutting motion by rotating the work-piece about the center axis of a spindle? This is the question that will be explored in this section.

6.3.1 Synchronisation of Spindle Rotation

Under typical operation of a DTM machine, the rotational position (θ) of the spindle is not controllable (spindle can only be programmed to rotate at complete revolutions at a certain rate). Also, the tool feed motion that happens in the XZ plane (Figure 6.2) is not related to the spindle rotation about the Z-axis. If tool motion, in the XZ plane, is represented as a function of time as $XZ(t)$ and spindle motion as $\theta(t)$, then under usual conditions, $XZ(t) \neq f(\theta(t))$. There is a mild form of synchronisation that happens, when the spindle speed is altered based on the X-position of the tool, to maintain constant surface cutting speed – however, this form of synchronisation does not involve controllable rotational position of the spindle; it merely changes spindle rotational speed with time based on where the tool is.

It is, however, possible to control the rotational position of the spindle as an additional axis of control in the DTM machine – so now we have a 3-axis simultaneously controllable DTM machine with X axis, Z axis and C axis (rotation around the Z axis – this is the normal spindle rotation, but now the angular position is known and controllable). We now have the situation where in $XZ(t) = f(\theta(t))$. To explain this idea further, consider the schematic shown in Figure 6.4. When the spindle rotational position changes from a position θ_1 to θ_2, during the normal rotational speed of the spindle to provide cutting motion, the position of the tool can be simultaneously controlled to move from $(x_1,0,z_1)$ to $(x_2,0,z_2)$ in a coordinated fashion. Such simultaneous control of θ and $(x,0,z)$ is required to create rotationally asymmetrical shapes. Hence, the variable θ is very much in the picture and cannot be set to zero.

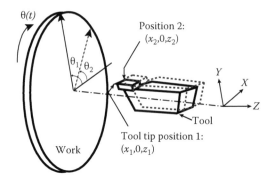

FIGURE 6.4
Schematic illustrating spindle synchronisation with X and Z axes. As the spindle rotates from angle θ_1 to θ_2, simultaneously the tool position can be changed from $(x_1,0,z_1)$ to $(x_2,0,z_2)$ synchronously.

This synchronisation of the spindle rotational axis, θ, with the tool motion (x and z) can be undertaken in two different methods. One method is by using the existing Z-axis motion slide in the DTM machine and moving the entire slide (on which the tool is mounted) in sync with the spindle rotation. The second method is by having an additional motion (W-axis) for the cutting tool over and above the existing Z-axis movement; either the W-axis alone is synchronised with the C-axis or the W-axis in combination with the Z-axis is coordinated with the C-axis. If the entire Z-axis slide is synchronised, the method is called slow tool servo, while coordination of a high-speed short-stroke W-axis is called fast tool servo. The slow tool servo is suitable for creating rotationally asymmetric features such as in Figure 6.1b while the fast tool servo is needed to make micro-features such as in Figure 6.1c. These two methods are explained in the following sections. New hybrid methods are recently reported, wherein both the Z-axis slide and the fast tool servo are used together smartly.

6.3.2 Slow Tool Servo (STS)

As explained above, the slow tool servo (STS) concept uses the existing motion of the Z-axis slide to synchronise with the spindle rotation to create rotationally unsymmetrical features on the work surface. This method of surface generation can be used as long as the features on the surface have gradual surface changes. Consider a hypothetical surface as shown in Figure 6.5. The surface is a flat circular disc with a hemi-cylindrical protrusion, of radius R, as shown, with its axis lying along a diameter. The radius R is shown exaggerated for clarity; also, the cylinder usually blends to the disk

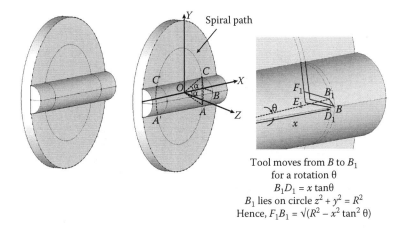

Tool moves from B to B_1
for a rotation θ
$B_1D_1 = x \tan\theta$
B_1 lies on circle $z^2 + y^2 = R^2$
Hence, $F_1B_1 = \sqrt{(R^2 - x^2 \tan^2 \theta)}$

FIGURE 6.5
Rotationally unsymmetrical (hypothetical) surface. The surface consists of a flat disc with a hemi-cylindrical protrusion running along a diameter. The radius of the protrusion is shown exaggerated for clarity; usually the radius is larger and blends with another concave fillet radius down to the disk.

surface with another concave fillet radius (not shown here). Clearly this feature is rotationally not symmetric.

To generate this surface, the tool again can be visualised (by stopping the spindle) to take a spiral motion path, but with a z-coordinate that is dependent on θ, the rotation about the Z-axis. Consider one segment of the spiral path shown in Figure 6.5. The origin is considered on the axis and at the center of the flat region of the disc. As the spindle rotates by an angle 2α from OC to OA, the tool has to undertake a helical path A–B–C. For the rest of the rotation along arc CC, the tool Z-position stays constant at $z = 0$. The tool then moves along helical path CA with varying z-coordinates and then again with constant $z = 0$ from A' back to A'' (it doesn't reach back to point A, but a different point A'' nearby – not shown in the figure – because of the spiral nature of the path). As the tool moves from point B to B_1 on the cylindrical surface (Figure 6.5), the coordinate changes can be written as follows in Equations 6.6 to 6.9:

$$x = x_{ref} - \frac{f}{2\pi}\theta; \tag{6.6}$$

$$\alpha = \tan^{-1}\frac{R}{x} \tag{6.7}$$

$$\text{for } x \geq R, Z = \begin{bmatrix} \sqrt{R^2 - x^2 \tan^2 \theta}, & \leq \theta \leq \alpha \\ 0, & \alpha \leq \theta \leq 180 - \alpha \\ \sqrt{R^2 - x^2 \tan^2 \theta}, & 180 - \alpha \leq \theta \leq 180 + \alpha \\ 0, & 180 + \alpha \leq \theta \leq 360 \end{bmatrix} \quad (6.8)$$

$$\text{for } 0 < x \leq R, \quad Z = \sqrt{R^2 - x^2 \tan^2 \theta} \quad (6.9)$$

where R is the radius of the cylindrical protrusion, and angle θ is set to zero when the cylinder axis coincides with the XZ plane during rotation. Obviously such motion is not possible without synchronising the spindle rotation and the Z-axis. The controller keeps track of the rotational angle and computes the x- and z-coordinates based on the above.

Earlier, there was a discussion of sampling points for non-analytical surfaces in symmetrical shapes. Such sampling points become applicable here for slow-tool servo motion based cutting tool path generation. As the spindle rotates, the z-direction motion of the cutting tool has to be controlled based on sampling points and interpolation between these points. There are two popular ways to choose these points as the spindle rotates (called azimuth sampling): constant angle sampling strategy (CASS) and constant-arc-length sampling strategy (CLSS) (see Figure 6.6). In CASS, equal angular rotational increments are chosen and the intersection of the Archimedes spiral with these angles is taken as a sampling point. As is evident from the figure, this results in crowding of sampling points closer to the center and sparser points farther from the centre, the density of points being dependent on radius. Choosing a finer angle, in order to achieve a reasonable error

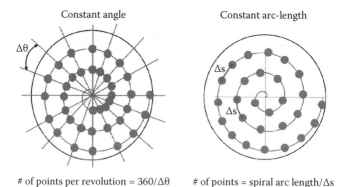

FIGURE 6.6
Two types of azimuth sampling methods.

in interpolation between points at the periphery, results in a large number of redundant points close to the center. While being easier to implement, the CASS results in a large volume of point-data for large optics (large enough to impede CNC controllers from handling the data-traffic) and more points per spindle revolution, leading to slowing down of spindle speeds. One way to overcome this problem is the constant arc method, which involves choosing samples based on equal arc-length spacing of points along the spiral path.

Implementing this constant arc method involves some complications, since the number of points per spindle revolution vary depending on where the tool is. New methods of choosing sampling points have evolved including combinations of CASS and CLSS [43]. These strategies of sampling points and interpolations influence the surface accuracy in slow tool servo systems and more so in fast tool servo systems (explained later in this chapter).

6.4 Tool Paths for Producing Micro-Features

As we have seen thus far, the slow tool servo system allows axially unsymmetrical features to be generated even while having a cutting motion that is rotational about a fixed axis. However, there are some limitations of the features that can be generated by the system. These limitations are due to the fact that the entire Z-axis slide system has to move synchronously with the rotating spindle C-axis. The Z-axis slide system consists of the table, the tool fixture and the tool itself, thus having a considerable mass in total. It is thus not possible for the Z-axis drive to execute, often back and forth, quick motions of the Z-axis system. Such quick motions are often required when the feature to be generated has steep slopes or the features are very small in size (micro-features). This is illustrated in Figure 6.7. The left feature of the figure is larger in size and has gradual slopes; hence, the change in z-coordinate of the tool as the tool traverses the red spiral path from A to B is small, as the spindle rotates by the angle shown. In contrast, the feature on the right side is small in size and has sharper slopes; as a result the change in z-coordinate as the tool traverses the spiral path from A to B is much sharper and happens over a narrow rotational angle of the spindle.

To achieve such quick back and forth motions of the tool in the Z-direction during a short quick rotation of the spindle, it would be better to have a separate actuator that will move the cutting tool alone (and hence a smaller mass to move), and not the entire Z-axis slide system. This system, which has an additional small parallel motion along the Z-axis (conventionally named W-axis) and being synchronously controlled with the C-axis, is called the fast tool servo (FTS) system.

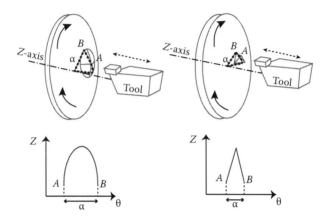

FIGURE 6.7
Small micro-features, such as in the right, to be generated by synchronous motion of the Z-axis and C-axis require quick back and forth motions along the Z-axis over shorter angular swings, when compared to large small-slop features such as in the left.

6.4.1 Fast Tool Servo (FTS)

The FTS system, often sold commercially as an attachment to the DTM machine, allows rapid acceleration and deceleration of the cutting tool in the Z-axis to move synchronously to the fast rotating spindle C-axis that provides the needed cutting motion. Such rapid motions are commonly achieved with a piezo-actuated linear actuator system on which the cutting tool and holder can be mounted (Figure 6.8).

A typical commercial system has a W-axis travel of several hundreds of micrometers with a frequency of about 200 Hz. This means that in a time of 1/200 or about 5 milliseconds, the system can travel, say, 300 micrometres along the Z-axis. If the spindle rotational speed is, say, 1000 rpm, this means that, by the time the spindle rotates by 6° (= 360/60 ms) the tool can be retracted and forwarded by 300 micrometres. Thus, very sharp slopes in

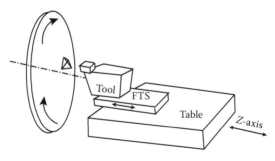

FIGURE 6.8
FTS in a DTM machine setup.

features can be generated. The FTS system has been very successful in generating, in a DTM machine, complicated micro-array features for use in varied optical applications.

Newer strategies that combine FTS and STS systems [44] and those with long-stroke, multi-axis adaptations of the FTS (explained later in Chapter 9) have also recently evolved.

6.5 Tool Normal Motion Path

As explained in Chapter 4, the cutting tool edge profile and its uniformity are critical to maintain the machined surface geometry. It is also mentioned that cutting tools can be commercially purchased with controlled and uncontrolled waviness of the cutting edge. Uncontrolled waviness tools have varying errors in edge-profiles along their edge length. In addition, tool wear alters the cutting edge profile and hence causes deviation from the planned machined surface profile. Additional axes of motion, for example, rotational motion about the Y-axis – the B-axis, can be utilised to reduce errors caused by such cutting edge profile changes. The so-called tool-normal cutting motion ensures that the same region of the cutting edge maintains contact over varying surface curvature of the work surface in the entire tool motion path (Figure 6.9).

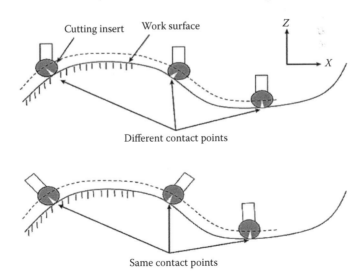

FIGURE 6.9
Ensuring that the same region of the cutting edge maintains contact, throughout the motion path.

The tool normal motion requires that the *B*-axis (on which the tool is mounted) be synchronised with the *X*-axis and *Z*-axis so that while the tool moves in the *Z*-plane, the surface normal can be tracked and the *B*-axis aligns the tool orientation along the surface normal. This way the same region of the cutting tool makes contact throughout the tool motion path. This has the following advantages:

- In case of uncontrolled waviness tool, a consistent area of cutting region generates the work surface; this ensures that the error variation along the cutting edge does not affect the machined surface. This error, being fixed, can perhaps even be compensated for;
- In case of controlled waviness tool, having a *B*-axis rotation helps in utilising different regions of the cutting edge and hence spread the wear evenly leading to longer tool life.

6.6 Deterministic Surface Generation

All mechanical machining, grinding and polishing processes, whether using single point cutting tools or abrasives, can be commonly understood as removing material by scratching action. Such scratching action requires moving the tool/abrasive or the work-piece in a certain direction at a certain speed. In machining and most grinding processes, this motion occurs along a single known direction in a repeated fashion – the cutting speed direction, often achieved via rotation. This results in scratches on the surface (when magnified enough for observation) with a known pattern – often called the lay pattern. Surface grinding results in straight line grooves/scratches on the surface, while face turning (the most common DTM operation) results in a spiral scratch pattern (Figure 6.10). This latter spiral pattern is consistent with the spiral tool motion path discussed in this chapter. The presence of

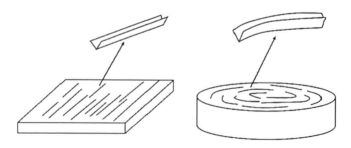

FIGURE 6.10
Lay patterns caused by deterministic tool motion paths. The patterns are scratch marks caused by abrasive/tool motion paths.

such lay patterns on the surface of a component is a result of what is called deterministic surface generation – a direct effect of the tool motion paths. That is, the tool motion path is deterministic and known a priori. Such lay patterns cause the surface to be anisotropic; in other words, the surface is not the same for a tiny person standing on it and looking at it in different directions – the topology appears different in different directions.

Such patterns often result in surface frequencies that are detrimental to performance in many applications such as visible optics. In other optical applications (Say IR Optics), these patterns may not severely affect the system performance. This interference with optical performance is a result of the interaction of wavefronts with various spatial frequencies present in the surface with which it is interacting. Surface roughness is generally measured by a surface profiler. If the spatial data can be collected with sufficient closeness, it is possible to obtain spatial frequencies present on the surface. Such frequencies are obtained using a Fourier analysis of the surface to split out the various frequencies in the surface profile signal, and also analyzing the power of each frequency – power spectral density (PSD). For example, a flat optical component DTM-machined at feed rate of 4 micrometers per revolution will result in a PSD peak at a spatial frequency of $1000/4 = 250$ mm^{-1}. The frequency components on a surface can be split into low spatial frequency (LSF), mid-spatial frequency (MSF) and high spatial frequency (HSF). The parameters quantifying LSF are form, power and figure, while HSF parameters are the commonly known roughness. MSF involves parameters such as waviness, ripple, smoothness and slope errors (Figure 6.11). The presence of

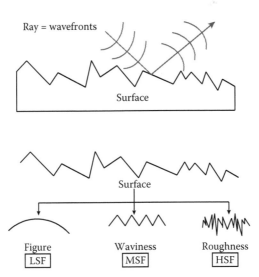

FIGURE 6.11
Deterministic tool motion paths discussed in this chapter often result in mid-spatial frequencies (MSF) on the surfaces produced.

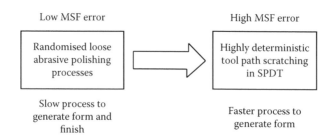

FIGURE 6.12
Low and high MSF errors are related to the type of ultra-precision machining processes.

lay patterns often results in these mid-spatial frequencies. It is well known in optics that the presence of MSF in surfaces results in effects like hazing, subtle reduction of the range of shades in dark areas of an image, or uneven texture in out-of-focus portions of an image.

The area under MSF errors is a consequence of the quick generation of the desired form on the work-piece. MSF related errors can be minimised using arbitrary motion paths, while scratching the surface for material removal. Such arbitrary motions can be achieved in loose slurry abrasive polishing processes (Figure 6.12). Such loose abrasive processes generate form very slowly, but can naturally get rid of MSF on the surfaces. Thus, visible optics components are often polished post-DTM, while IR-optical components (if the tolerances are generous) can be used without such further processing steps. The fast tool servo systems discussed earlier are modified to provide random motions superimposed on the regular slide motions in an effort to remove the deterministic marks and hence to reduce the influence of MSF on optical performance.

6.7 Summary

This chapter presents the tool motion paths often used in the DTM process to generate the various optical surfaces. The face-turning process is the often-used process in DTM and the tool motion path can be visualised as an Archimedes spiral motion, leading to typically axially symmetric features. By using special synchronising of axes movement, it is possible to generate axially non-symmetric features (both large gradual-slope and small micro steep-slope types) with the STS and FTS systems. The chapter also showed that the tool motion paths used in DTM naturally result in deterministic surfaces with mid-spatial frequency components in the surfaces. FTS systems can be used innovatively to try to minimise such errors. However, often subsequent polishing processing is required, at least for visual optics components.

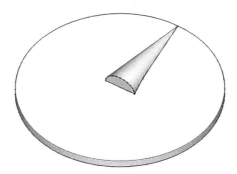

FIGURE 6.13
Figure for Question 5.

6.8 Questions and Problems

1. Explain and differentiate slow tool servo and fast tool servo systems in a diamond-turn machine.

2. Why is spindle-axis synchronised movement needed to generate rotationally asymmetric parts in a diamond-turn machine?

3. Describe the various spatial frequencies on a surface and their consequences on optical performance.

4. Correlate MSF and tool motion paths.

5. A hypothetical surface to be generated using a slow tool servo system is shown in Figure 6.13. The surface is a cylindrical disk with a right-angled half-cone protrusion lying with its axis along one of the radii. Write the generic coordinate variation for X and Z as a function of rotational angle θ for a finish turning pass on this surface.

7

Application of DTM Products

7.1 Introduction

Ultra-precision machining has become an enabling tool to achieve very high levels of performance in various application areas. Large-sized optical elements like mirrors and lenses for application in astronomy were earlier manufactured manually and took enormous amounts of time and skill. Subsequently, precision machining was used to shape them and manual polishing or abrasive-based finishing processes were used to improve their accuracy level as well as cycle time. With the advancement in ultra-precision machining techniques like the development of diamond turn machines, the finishing process has become deterministic in nature and led to improved quality and reduced cycle time. Emergence of biomedical devices has opened up a newer opportunity for diamond turn machining. Similarly, the need for many ultra-precision components in various fields, including aerospace and military equipment, necessitated the use of diamond turn machining and helped in developing various configurations of machines to generate complex surfaces with very high levels of accuracy. Use of extensive polymer optics in day-to-day use as well as for emerging advanced applications is possible, owing to the developments in precision machining equipment, like diamond turn machines. This chapter presents a detailed discussion to give an overall idea about the application of diamond turn machining in various fields.

7.2 Diamond Turn Machining Applications

Diamond turn machining caters to a wide range of application areas. Major groups of components that can be machined by DTM are categorised as follows:

- Polymer optics
- Metal molds for polymer optics
- Metal optics (electro optics)
- Ultra-precision components

Each one of the above groups of components further consists of many diversified types of components. The optics, both polymer and metal, constitute a major portion of diamond turn machined components. Small batch production of polymer optics is directly machined on DTM whereas metal molds machined by DTM are used to achieve high volume production of polymer optics; in the case of metal optics, only the machining technique is used and not molding. Generally, diamond turn machining is extensively used for machining UV, broadband and IR optics. In addition to the optical domain, many ultra-precision components, where size control is the primary requirement in addition to the control on shape error and surface finish, are finished using diamond turn machining.

7.3 Applications in the Optical Domain

Optics can be classified based on their material of construction, type of light behaviour on the optical surface and wavelength domain. Figure 7.1 shows the classifications of the optics that can be manufactured using diamond turn machining. Figure 7.2 shows the basic optical elements that can be machined by diamond turn machining. Table 7.1 summarises their characteristics and the relevant diamond turn machining facility required to manufacture them.

The following equation describes the functional relationship between optical scattering and surface roughness [45]:

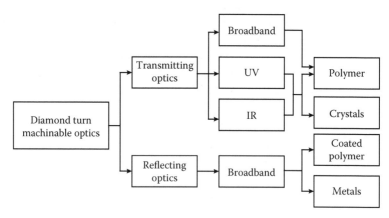

FIGURE 7.1
Diamond turn machinable optics.

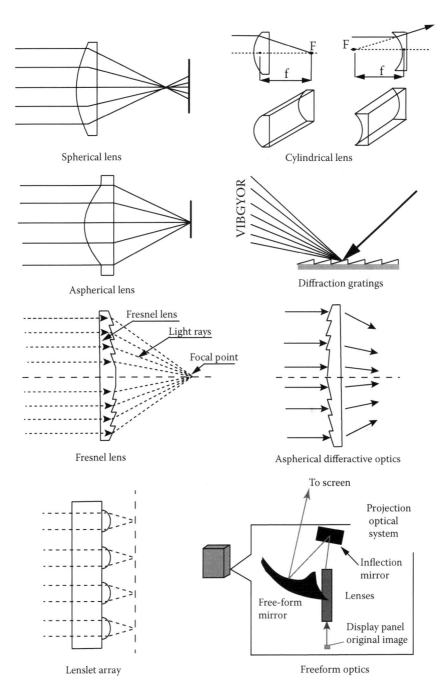

FIGURE 7.2
Basic optical elements.

TABLE 7.1

Diamond Turn Machineable Optical Elements

Facility	Characteristic (typical application) [critical parametres]	Facility Required
Spherical lens	Focuses light into a point (projection, collimation, imaging, ophthalmic) [radius of curvature, shape error]	2-axis DTM
Cylindrical lens	Focuses light into a line (beam shaping and ophthalmic) [radius of curvature, shape error]	2-axis DTM
Aspherical lens	Reduces or eliminate spherical aberration and astigmatism (ophthalmic, projection, collimation, imaging)	2-axis DTM
Diffraction gratings (reflection)	It has periodic structures like grooves to diffract light into several beams travelling in different directions (beam splitter, beam shaper, hologram) [groove shape accuracy, spacing, depth]	2-axis DTM With fly cutting
Fresnel lens (transmission)	Reduces spherical abrasions (collimation, e.g. light house, collection, e.g. solar collector, low cost magnification) [focal distance, diameter, prism width, wedge angle, collecting angle]	2-axis DTM
Aspherical diffractive optics	It has diffractive features on aspheric surface. Removes spherical and chromatic abrasions (imaging, broadband illumination sources) [sag at any radial distance, conic constant, curvature, aspheric terms, groove shape accuracy, spacing, depth]	2-axis DTM
Lenslet array	Concave or convex spherical or aspheric micro lenses in a plane (Shack–Hartmann sensor, beam homogenisation for projection systems) [aspheric parametres, pitch]	3-axis DTM Fast tool servo
Freeform optics	Non-symmetric surface forms (LED reflector, HUD, progressive spectacle, torics, off-axis parabola) [free form shape parametres]	3-axis DTM and slow tool servo (or) 4-axis DTM with fly cutting

$$TIS_{BP}(R_q) = R_0 \left[1 - e^{-\left(\frac{4\pi R_q \cos\theta_i}{\lambda} \right)^2} \right] \qquad (7.1)$$

In this equation, TIS_{BP} is total integrated scattering; R_0 is the theoretical reflectance of the surface; R_q is the RMS roughness of the surface; θ_i is the angle of incidence on the surface; and λ is the wavelength of light.

For a given surface roughness value, scatter increases with the decreasing wavelength. Therefore, surface roughness value of the optical surface for the applications in the visible range (λ = 400 to 700 nm) should be less compared to the applications in the IR range (λ = 700 nm to 1 mm) and further smaller for the applications in the UV range (λ = 10 nm to 400 nm). For example, surfaces used for 193-nm wavelength applications should be an order of magnitude smoother than surfaces used for visible applications to achieve the same scatter specification. Diamond turn machining is capable of generating surface roughness values of better than 5 nm and hence it readily meets both visible and IR requirements; for UV application, further super finishing may be required to reduce the scattering.

7.4 Polymer Optics Products

Polymer optics include LED optics, light reflectors, freeform optics, infrared optics, imaging optics, display optics, street light illumination, IR and UV optics and ophthalmic lenses. Figure 7.3a and b show some typical polymer optics.

7.5 Mold Inserts for Polymer Optics

Metal molds are used in injection molding to mass manufacture polymer optics. Due to high quality components on the optical products, the manufacturing process of the molds has to be accurate and precise. Thus, the master mold needs to be manufactured to very high precision, in terms of size, shape accuracy and surface finish. Electroless nickel (with a higher percentage of phosphor) is coated (thickness of 250–500 microns) on steel/stainless steel substrates. These substrates are then diamond turned for

(a)   (b)

FIGURE 7.3
Typical polymer optics products. (a) Typical small size polymer optics. (b) Typical large size polymer optics.

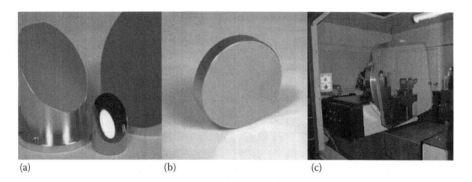

FIGURE 7.4
Typical metal optics. (a) Typical off-axis parabolic mirrors. (b) Typical planner aluminium alloy mirror. (c) Large size aluminium alloy mirror for space application.

use as mold inserts. Various shapes of optical components specified in Table 7.1 are produced using metal molds. To fabricate complex mold inserts by DTM, various accessories like slow tool servo, fast tool servo etc are used.

7.6 Metal Optics

Metal optics is another major area where DTM is extensively used. Reflective types of mirrors for various applications ranging from laser to astronomy are machined from materials like copper and aluminum alloy. Generally, these mirrors are conic sections in geometry, requiring stringent control on shape and surface finish. These are machined by 2- or 3-axis diamond turn machines. Figure 7.4a, b and c shows some metal mirrors.

7.7 IR Optics

Infrared and near infrared optics are extensively used for night vision and thermal imaging applications in various fields including defence, aerospace, automotive and medical. They are generally of transmission type. However, in some applications, reflective IR optics is also used. The majority of these optics are manufactured from semiconductor materials like silicon and germanium and a few are from glass and polymers. Figure 7.5a, b and c shows some IR optics.

(a)   (b)   (c)

FIGURE 7.5
Typical IR optics. (a) Typical germanium optics. (b) Typical silicon optics. (c) Typical silicon diffractive optics.

7.8 Diamond Turn Machined Ultra-Precision Components

Precision optics need strict form, figure and finish control. However, many other engineering products require size control as a primary requirement. Also, many engineering products require shape and size control, for example, drum for flat screen TV panel and high precision ball bearing. Figure 7.6a, b, and c shows some engineering products finished by diamond turn machining.

7.9 Major Diamond Turn Machining Application Areas

Diamond turn machining is extensively used in three major areas, namely, aerospace and defence, industrial and biomedical applications [46,47]. Table 7.2 shows the major areas and some specific applications.

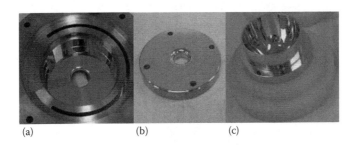

(a)   (b)   (c)

FIGURE 7.6
Typical diamond turn machined industrial products. (a) Typical linac cavity. (b) Typical flat precision spacer. (c) Typical light concentrator with flange.

TABLE 7.2

Application Areas of Diamond Turn Machining

Aerospace and Defence	Industrial	Biomedical
Missile system	Laser	Dental imaging
Heads up display (HUD)	Heads up display (HUD)	Medical imaging
Helmet mounted display (HMD)	Automotive	Molecular imaging
Communication	Optical sensors	Illumination
Security	Display and projections	Medical laser devices
Biometrics	Solar energy	Hip joint

Most popular products of DTM include HUD lenses, astronomy mirrors and lenses, high precision image intensifier tube, intraocular cataract lens (IOL), low vision aids, mobile camera lenses, night vision goggles, progressive lenses, orthopedic ball-socket joints, hip joints, knee joints and nuclear fusion components.

7.10 Materials Machinable by DTM

Diamond turn machining can be carried out on metals, polymers and crystals. These materials can be finish machined to have the required surface finish and shape without causing rapid tool wear. Some of the most extensively used materials are listed below.

7.10.1 Metals

Metals that can be diamond turn machined are

> Copper, brass, aluminum alloys, electroless nickel, bronze, copper beryllium, tin, antimony, silver, gold, zinc, magnesium, lead and platinum.

7.10.2 Polymers

Polymers that can be diamond turn machined are [48–50]

> PMMA (Chemical & Scratch resistance), Polycarbonate (Impact strength/temperature resistance), Polystyrene (Low cost/highly transparent), PolyEthirmide (High Thermal/chemical/Impact resistance/high index), Cyclic Olefin Copolymer (High modulus/low moisture absorption), Cyclic Olefin Polymer (Completely Amorphous/high chemical resistance), Nylon, Acrylonitrile Butadiene Styrene.

7.10.3 Crystals

Crystals that can be diamond turn machined are [48,49]

Barium Fluoride (BaF_2), Cadmium Telluride (CdTe), Cesium Bromide (CsBr), Cesium Iodide (CsI), Chalcogenide Glass, Gallium Arsenide (GaAs), Germanium (Ge), Lithium Fluoride (LiF), Magnesium Fluoride (MgF_2), Potassium Bromide (KBr), Potassium Chloride (KCl), Silicon (Si), Sodium Chloride (NaCl), Thallium Bromoiodide (KRS-5), Zinc Selenide (ZnSe), Zinc Sulfide.

7.11 Summary

This chapter discusses various applications of diamond turn machining. The relevance of DTM for machining of optics is highlighted. Applications include polymer optics and mold insert, metal optics and other ultra-precision engineering products. Different types of optical elements and various metals, polymers and crystals that can be diamond turned are listed.

8

DTM Surfaces – Metrology – Characterization

8.1 Introduction

Precision engineering has two inseparable dimensions of material processing, namely, deterministic fabrication and error free metrology. Diamond turn machining (DTM), a specific branch of precision engineering, amply demonstrates the need to qualify the fabricated component for its adherence to both dimensions and surface quality within prescribed tolerance ranges.

A major branch of qualification of DTM-generated components involves surface metrology and surface characterization. Often these two terms, metrology and characterization, are used without differentiation in between [51]. However, it is pertinent not to complicate these issues. This can ably be done by a comprehensive discussion and clear understanding of the surface features as per desired quality criteria.

Metrology refers to just measurement of the geometrical features and surface features of the component fabricated. Characterization refers to a holistic approach of assessing the features' departures from the specifications, analysing them in relation with each other, with inputs for their possible reduction by process optimisation. Figure 8.1 represents the relevance of qualification of precision surface quality, in terms of geometrical specification and surface features in the cycle of precision component development by DTM [52].

DTM allows high precision surfaces to be manufactured quickly and efficiently. As with any precision machining processes, the outcome of the DTM also needs to be qualified and quantified.

The precision surfaces generated by DTM are generally assessed for their dimensional accuracies (whether they met the specified geometrical dimensions within the prescribed tolerances) and for their surface quality criteria, to be introduced later in this chapter. Figure 8.2 presents the typical evaluation methodology of the DTM-generated precision components.

Apart from the departures from the geometrical dimensions (as indicated in Figure 8.2), the features of DTM-generated precision components and

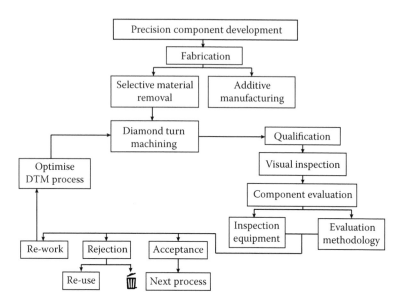

FIGURE 8.1
Precision component production cycle.

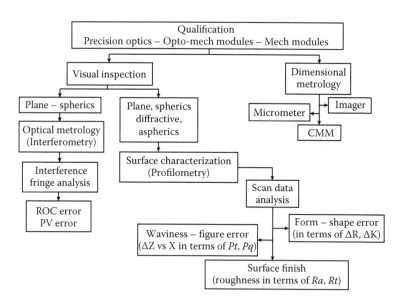

FIGURE 8.2
Evaluation outline of DTM-produced precision components and surfaces.

surfaces depart from the prescribed surface parametres [52] in terms of *form, figure* and *finish* quality criteria (Figure 8.3).

The smoothness/roughness is the primary surface feature, which can be sensed by a scanning probe. This is followed by the analysis of departure of the surface's (fabricated) locus (from its prescribed locus), elucidating the shape and *figure* errors.

A majority of global DTM operations involve the generation of rotationally symmetric surfaces, by turning/facing operations (Figure 8.4). DTM processes are known to deliver near-accurate surfaces (as per prescribed nominal values) with a high degree of precision. This helps in maintaining the surface quality well within the prescribed tolerances. However, due to a multitude of factors, DTM processes have significant inherent processing errors. Major issues that severely affect the surface quality include: (1) material

FIGURE 8.3
The curved line shows the desired profile; the double dimpled curve shows the waviness; the second double dimpled curve shows the waviness with roughness spikes.

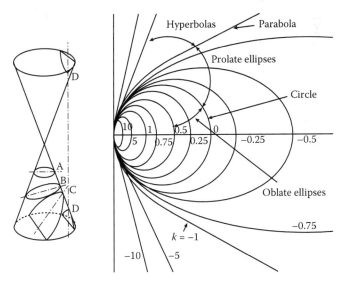

FIGURE 8.4
DTM turnable precision conics. A, spherical; B, elliptic; C, parabolic; and D, hyperbolic profiles.

(of work-piece) characteristics; (2) machining parametres; (3) tool geometry and tool placement; and (4) machining and metrology protocols. The discussion on these factors merits greater mention elsewhere in this book. This chapter focuses mainly on the measurement issues of the DTM-generated surfaces.

8.2 Surface Quality

The discussion of the study of surface errors needs to be both qualitative and quantitative. A majority of the precision surfaces generated by DTM processes for a multitude of applications in numerous domains of precision instrumentation are: plane, spherical and conics. The geometry and surface features are formulated by the designer for the intended performance of the component (both as component and as a module of the whole system). However, any variations in its geometry and profile features may severely affect its anticipated performance. Consequently, it is necessary to understand both (a) the dimensional parametres and the surface features of the precision surfaces and (b) the errors therein, generated due to the factors mentioned above.

The DTM-generated precision conic surfaces are defined in terms of their radius of curvature, conic constant and the useful aperture of the surface. This surface is given in terms of its sag (z) versus useful aperture (r):

$$z = \frac{cr^2}{1+\sqrt{1-(1+k)c^2r^2}} + \alpha_1 r^2 + \alpha_2 r^4 + \alpha_3 r^6 + \alpha_4 r^8 + \alpha_5 r^{10} + \alpha_6 r^{12} + \alpha_7 r^{14} + \alpha_8 r^{16}$$

$$(8.1)$$

where, c = Base curvature of the surface (inverse of radius of curvature) at the vertex of the surface; k = Conic constant; $k > 0$ (oblate ellipsoid, rotating on its major axis); $-1 < k < 0$ (prolate ellipsoid, rotating on its minor axis); 0 (sphere); -1 (parabola), < -1 (hyperbola); while A_i corresponds to the constants of higher order conic terms.

8.2.1 Form Error

The conic is formed by the locus of the tool interaction on the work-piece around its rotational axis, following its base radius of curvature (RoC) and the value of conic constant k. In the DTM of this locus, what can go wrong? Without counting the practicality of conic surface generation issues, anything can go wrong (even) theoretically! Either the nominated base radius or the desired conic constant or both may change in the machined surface. The outcome may be altogether another surface or the desired surface with gross profile error popularly known as *form error*. The genesis of this *form error* and its elimination/reduction is discussed elsewhere in this book.

8.2.2 Figure Error

By optimising the DTM process, one may maintain the nominated values of RoC and k (within the given tolerance ranges) and eliminate or reduce (significantly) the *form error*. Still, the surface profile may depart from the desired locus in terms of 'number of peaks and valleys', with varying amplitudes and spatial distributions, throughout its useful aperture. These departures in terms of 'peaks and valleys' will also affect the surface quality and the intended performance significantly. It is common amongst DTM practitioners to club these two surface errors, namely, *form error* and *figure error*, into a generic profile error, often known as 'waviness'.

8.2.3 Finish Error

The DTM process, due to its controlled machining forces and reduced vibrations etc., offers the best surface smoothness compared to any other physical material removal process. Despite these advantages, the DTM-generated precision surfaces still show (though minimal) some roughness. The smoothness of the machined surface is expressed in terms of its residual roughness [53]. This presence of roughness of the surface is known as *finish error*. Roughness is due to the irregularities, which are inherent in the production process (e.g. cutting tool and feed rates). The roughness also depends on the material composition and heat treatment.

In DTM-generated surface texture evaluation, the basic discussion starts with differentiating between the *form error*, *figure error* and *finish error*. *Form error* is the general shape deviation of the surface from the intended shape. *Figure error* corresponds to the case, where the surface has peaks and valleys from the datum, though the basic shape is retained. The roughness corresponds to the localised amplitude variations from the peaks/valleys RMS locus in the prescribed sample length. It is also necessary for the single-point diamond turning teams to understand at what point waviness gets reduced and *finish error* begins. Subsequently, it is vital to comprehend when the roughness starts impacting the waviness.

8.3 Quantification of Surface Errors

It is one thing to identify the surface errors distinctly and it is quite another thing to quantify them, in universally acceptable and understandable terms, without any scope of error in the process of their derivation, analysis and optimisation.

This calls for a standardised approach to express the surface quality. Accordingly, a whole new range of parametres are introduced with clear definitions (without any overlap) and scope. Here it is apt to introduce surface texture parametres, which express surface quality both in terms of waviness and roughness.

8.4 Surface Texture

The discussion of classifying and quantifying various parametres of surface characteristics need to be preceded by some understanding of surface texture. What is surface texture? Based on the machining method deployed, material machined and its final intended application, the machined surface will have a few characteristics in terms of deviations from the intended surface profile in varying degrees (Figure 8.5). The deviation profile (with changing frequencies and magnitudes) constitutes the actual 'lay' of the machined surface and is known as the surface texture [54]. This comprehensive term includes the form error, figure error and finish error (as defined earlier). Once the surface texture is understood qualitatively, the discussion calls for quantification of the comprehensive surface profile error in its various constituents. For this to happen, we need to define and explain various parametres of surface texture.

The science of surface characterization is a complex fusion of multiple disciplines. Though its relevance is well understood, its methodology is sadly ignored by the user community. However, it is essential to understand that a few basic questions need to be asked and answered before a surface is gainfully characterized, namely: (1) machining process by which the precision surface is developed; (2) where (geographical location) this surface is

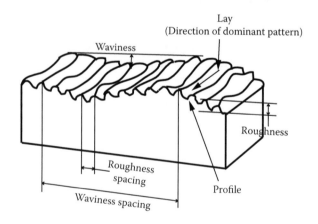

FIGURE 8.5
3-D perspective of DTM surface with lay, waviness and roughness.

intended to be deployed; (3) whether it will be used in conjunction with other precision surfaces; (4) what is the application area of this precision surface; and finally, (5) what is the process etc.

The characterization methodology and the relevant equipment can primarily be decided by the answers to the first three queries. In earlier days of surface characterization, location of the activity used to decide on the type of surface parametre for probe. For example, in the United States, average roughness (Ra) was quite popular; while in Europe, mean roughness depth (Rz) was more prevalent. Nevertheless, in today's scenario, with precision products being developed, distributed and used on a global basis, geographical considerations have lost their relevance. However, this does not mean that the quality criteria are compromised, only some of the surface texture parametres became popular, while some have become irrelevant. The quest for better surfaces is still on and will only intensify in coming days.

There is a plethora of equipment available to characterize precision surfaces fabricated by various precision fabrication techniques. However, our discussion in this context is limited to only diamond turning. Though a majority of DTM-generated surfaces are deployed in optical applications (using general conic surfaces for imaging and non-imaging), other application sectors include consumer instrumentation, security equipment, biomedical instrumentation, automobile instrumentation, societal instrumentation, avionics and strategic sectors. Almost all the diamond turned surfaces are characterized for their surface texture attributes by using a contact profilometer.

The contact profilometer (Figure 8.6) provides a linear scan of the machined surface through its stylus (of pre-designated geometry). This surface

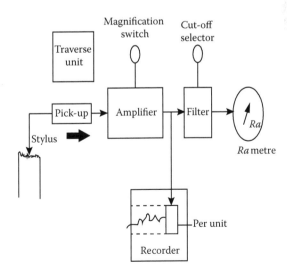

FIGURE 8.6
Schematic of stylus-based contact profilometre.

scanning is followed by the comparison of the scan with the target/intended surface profile, and finally by the analysis of the resultant mismatch (namely, profile errors) of the textures of desired and developed surfaces. In a precision machined surface of a designated shape, namely, plane, rotationally symmetric conics (spherical, ellipsoid, parabolic and hyperbolic profiles), toric or freeform, the profile of the surface is measured in terms of depth (sagitta) value versus the aperture value. In a contact profilometer, a diamond/ruby stylus extended from a scanning arm moves over a precision machined surface in a linear mode, and the imperfections of the machined surface result in vertical stylus displacements as functions of the positions of the stylus. The comparison of the designated sagitta versus the actual sagitta with respect to the aperture provides us with the variation of the profile. The resultant error profile is the manifestation of the departure of the profile from its designated one.

A properly selected combination of (1) stylus geometry, (2) stylus scan logistics, (3) matching electronics of profilometer, (4) standard operational protocols of metrology, and (5) appropriate surface analysis software will help in computing the surface texture of a precision machined surface quality with high degree of fidelity. The contact profilometry has some notable advantages of surface metrology, while its demerits also needs to be considered for a holistic estimate of surface characterization. More of this, may be later.

8.5 Surface Texture Parametres

The surface texture of any precision machined component can be defined into two broad constituent components in terms of their spatial wavelengths [55–59].

The longer wavelengths form the 'waviness' realm – caused by macro-type influences – while the smaller wavelengths are included in the domain of 'roughness' – triggered by the tooling and machining process. It is to be noted that both waviness and roughness can be controlled by the condition and quality of the tooling. Both can be managed by operator's skills and functional requirements of the components. Apart from the spatial wavelengths, the frequencies and the magnitudes of the deviations from the baseline also play a major role in the quantification of surface texture. A large number of surface texture parametres have been introduced by the precision machining and metrology community. There are more than 100 parametres, which are used to express the texture [60]. Putting a number to the surface texture is a task that needs to be handled with some caution and foresight. As mentioned earlier, the process of machining, the method of measurement and the application of the precision surface will decide on how to express the texture and which parametres need to be considered. Since these parametres bring all the surface data into a single value, great care must be taken in applying and interpreting them.

To understand the quantified nature of a precision machined surface texture, it is advisable to be familiarised with the meanings of the surface texture parametres to be introduced hereafter. The discussion on surface texture parametres should include the genesis, usage, popularity and prevalence of the terminology of surface parametres and attributes. Assuming that most of the DTM-processed surfaces are characterized by a stylus-based profilometer, we will discuss the surface characterization terminology in detail.

> *Surface:* The boundary of the processed work-piece medium, which separates from the surrounding medium.
>
> *Profile:* A two-dimensional slice through an area of the machined surface.
>
> *Surface texture:* The topography of a surface composed of certain deviations (including roughness and waviness as discussed earlier).
>
> *Parametres:* The features of the surface texture expressed in terms of either the occurrences of profile departures (from the locus of the profile) or in terms of their magnitudes.
>
> *Filter:* A procedure to select a particular wavelength (of surface profile deviations) range by leaving out the rest of the wavelength ranges of the profile deviations, in terms of peaks and valleys [61]. The surface profile characterization methodology involves a clearly defined filtration process of these peaks and valleys of given magnitudes and spatial distributions over the locus of the profile [54]. The software codes associated with the surface characterization equipment (profilometres) are exclusively developed for the filtration process (Figure 8.7).
>
> *Primary profile:* The primary profile is derived from the surface profile with the form (the shape of the surface) being removed. It includes all deviations from the locus of the designated profile, including large wavelengths (figure errors) and small wavelengths (roughness). The primary profile forms the basis to evaluate all primary parametres.
>
> *Waviness profile:* The waviness profile is derived from the surface profile, including all deviations with the large wavelengths (from the locus of the designated profile), while the departures with the small wavelengths (roughness) are not considered.

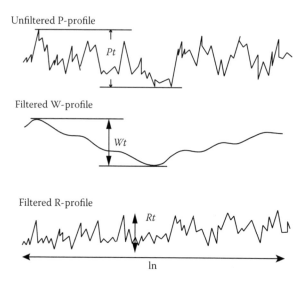

Unfiltered P-profile

Filtered W-profile

Filtered R-profile

FIGURE 8.7
Unfiltered and filtered profiles of surface texture.

> *Roughness:* A surface feature of process indentations left on the topography of the surface profile, due to machining process and other material characteristics. It is noted that this excludes waviness and form errors [53].

The unfiltered primary profile (P-profile) is the actual measured surface profile. Filtering it in accordance with ISO 11562/ISO 16610-21 produces the waviness profile (W-profile) and the roughness profile (R-profile). The variable for determining the limit between waviness and roughness is the cut-off λ_c. The profile type is identified by the capital letters P, R or W.

> *Areal:* A three-dimensional surface area of the surface under consideration.
>
> *Stylus instrument:* It enables two-dimensional tracing of a surface. The stylus is traversed normal to the surface at constant speed.
>
> *Traced profile:* The enveloping profile of the real surface acquired by means of a stylus instrument. The traced profile consists of form deviations, waviness and roughness components.

The surface scan has two parts of information: the profile length details and profile departures (in terms of waviness, roughness, etc.). In order to qualify the profile for its profile violations adequately, it is necessary to define the length parametres in terms of total scan length, evaluation length, sample length etc. (Figure 8.8).

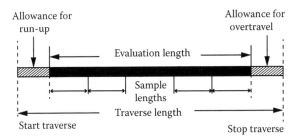

FIGURE 8.8
Scanning of surface length – length parametres.

Traversing length: It is the overall length traveled by the stylus, when acquiring the traced profile.

Evaluation length: The portion of the traversing length, which is considered for the analysis of the surface.

Sampling length: The portion of the evaluation length considered for assessment and evaluation of roughness (with pre-decided wavelength), also known as the cut-off length.

Cut-off: The variable determining the limit between waviness and roughness profiles, in terms of wavelength. This value is selected as per workpiece surface, based on either peak-valley distribution or as per expected roughness. It is vital to note that different cut-offs will give different values of the same surface texture (roughness). Also, determination of the appropriate cut-off is application specific. Proper cut-off lengths are to be decided by area of interaction with mating components, material properties, and the scale of physical phenomena and so on.

8.6 Spatial Parametres

A series of surface texture parametres are categorised under a broad category of *spatial parametres*. These parametres are defined on the evaluation length and are based on periodicity details of the reference line of the surface under consideration.

8.7 Amplitude Parametres

In consonance with the spatial parametres, another series of surface texture parametres are classified under the broad category of *amplitude parametres*, the amplitude in question being the quantum from the reference line. Here, the reference line needs an unambiguous definition. The spatial and amplitude parametres together define comprehensively the profile errors of the surface generated. It is necessary to understand that the type of texture parametre is decided by the profile deviation: waviness or roughness (Figure 8.3).

> *Average height parametre:* Surface texture expressed in terms of average height magnitudes' estimates computed by taking of means (arithmetic or RMS) of heights' estimates from all individual sampling lengths. To express roughness profile, five sampling lengths are considered for averaging.

> *Extreme height parametre:* Surface texture expressed in terms of maximum height magnitudes' estimates computed by taking the total amplitude of the highest peaks and the lowest valleys. This is a very significant parametre to define the quality of the precision surface generated in the machining process. In order to qualify the surface's suitability for the desired application, this parametre is generally considered for the total evaluation length rather than the sampling length. Many times, this parametre will decide the selection of the machining process in order to meet the prescribed surface quality tolerances, as well as the acceptance of the surface thus generated.

A multitude of surface texture parametres have evolved over the years across various working groups, often leading to greater confusion and incomprehension. Hence, it is advised that one needs to consider only those sets of surface texture defining parametres that meet the surfaces' stated application genre and the quality considerations.

In order to give a practical insight to DTM surface characterization, let us review the most commonly considered surface texture parametres: *Pt, Pq, Ra* and *Rt.*

> *Pt: Peak-to-valley profile amplitude error:* In most of the surface characteristic evaluation and acceptance/rejection analyses, the most significant surface quality parametre is *Pt*. This refers to the sum of the amplitudes of the highest peak and that of the deepest valley (of all the peaks and valleys resident) on the total evaluation length of the primary profile of the DTM surface. One of the major applications of DTM is precision optical surface generation (for a wide spectrum of instrumentation applications). In Figures 8.9 and 8.10, the profile analyses of two DTM surfaces are shown.

Figure 8.9 corresponds to a DTM generated surface (with 49 mm evaluation length) with a large value of *Pt*: 1.74 µm, while Figure 8.10 corresponds to a well-turned surface (with 47 mm evaluation length) with a relatively small value of *Pt*: 0.33 µm. The surface profile error *Pt* as profile-scanned in

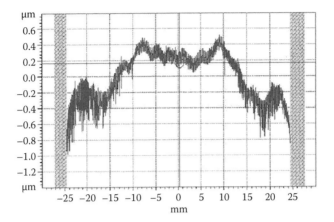

FIGURE 8.9
DTM surface with large *Pt* value: 1.74 μm and *Pq*: 0.3 μm over 49 mm evaluation length.

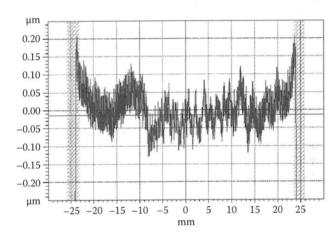

FIGURE 8.10
DTM surface with small *Pt* value: 0.33 μm and *Pq*: 43 nm over 47 mm evaluation length.

these profiles is analogous to the PV error used in optical metrology terminology. The surface profile error of an optical surface is generally expressed
in terms of the mean wavelength (of the visible part of the optical spectrum):
0.633 μm. Any good optical surface has a profile error of a fraction of this
mean wavelength. In the case of DTM surfaces denoted in Figures 8.9 and
8.10, the surface error is about 2.7 λ and 0.52 λ, respectively. The application areas will decide whether the surface is good for the intended purpose,
rather than its exact or empirical value. The *Pt* value decides the fate of the
optical component between acceptance and rejection, thereby the future of
the industry where the DTM components are developed.

Pq: Average amplitude error: In the family of waviness parametres, *Pq*, the root-mean-square profile error is the second most important surface quality parametre to assess the profile departures from the prescribed surface profile. In some of the optical design specification sheets, *Pq*, the rms profile error is also considered for qualifying the surface quality. However, it may be noted that, *Pq* is an average value of the total amplitudes (of peaks and valleys) and, hence, may not adequately represent the true nature of the surface texture. Also, this parametre may hide the process-induced surface errors on the profile during its averaging. Therefore, in order to validate the machining process as well as the machined component, it is advised to consider the largest possible surface profile error, i.e. *Pt*. Nevertheless, the intended application decides the usefulness of the rms profile error of a surface.

Ra: Average roughness error: The surface roughness is the most sought-out surface quality parametre by all the sects of precision machining fraternity, due to its utility, popularity and acceptability. Again, a large number of parametres have been devised to express the roughness features over the years, based on geographical, skill-sets available, applications intended and knowledge growth in the precision metrology community. It is advantageous to align the machining and metrology operations with a single parametre to express the surface quality, which is acceptable by all the concerned. The arithmetic average roughness parametre *Ra* broadly meets this purpose, under many application domains of precision engineering. *Ra* denotes the surface roughness value, corresponding to the arithmetic average of amplitude variations from the rms mean line of the roughness datum line, for a selected sample length of the DTM surface, over the entire evaluation length. Figure 8.11 shows the roughness profile of a DTM surface (over an evaluation length of 12.8 mm).

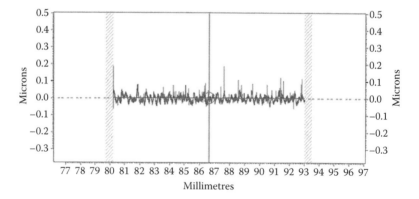

FIGURE 8.11
DTM surface with small *Ra*: 14 nm and large *Rt*: 248 nm over 12.8 mm evaluation length.

The selection of a sample length for the analysis of surface roughness varies from application to application and for different levels of operations as well. For example, the respective sample lengths for the R&D stage, prototype production and for large volume production vary drastically. The typical values are 0.025 mm, 0.08 mm and 0.8 mm, respectively. As in the case of the surface profile error (*Pt*), the roughness quality criterion of the DTM-generated surface is decided by its intended application.

> *Rt:* This denotes the surface roughness value corresponding to the sum of the amplitude of the highest peak and that of the deepest valley of the roughness profile, in a selected sample length of the DTM surface over the entire evaluation length.

Any discussion on surface characterization is not complete without a query of where roughness ends and waviness begins. Figure 8.11 provides a very interesting scenario. Here, the average roughness value is low (only 14 nm). Nevertheless, the peak roughness value is very high: 248 nm. This means that there are spikes of roughness with high value of *Rt* (248 nm) on the surface contrasting with the average roughness profile. These spikes might have been hidden in the amplitude averaging process (for *Ra*, the arithmetic average roughness parametre or for *Rq*, the rms average roughness parametre). Nevertheless, its presence will contribute to the waviness profile and may affect the *Pt* value significantly, if the roughness spikes happen to be on the cusp of the deciding peak/valley of the waviness profile. Hence, it is advisable to generate a DTM surface, with not only a good average roughness but also with a manageable *Rt* value, in order to restrain *Pt*.

8.8 Power Spectral Density

The roughness profile of a DTM-generated surface is generally expressed in terms of the arithmetic average values of amplitudes of the localised peaks and valleys in a given sample length. This approach neglects the spatial distribution of these peaks and valleys on the entire scan length (Figure 8.12). This may lead to a situation where different roughness profiles may show the same arithmetic value, while having different surface texture [62]. This presents a flawed picture of the DTM surface. Hence, a universally accepted surface roughness parametre, power spectral density (PSD) has evolved in the last few years [54,63].

This single parametre quantises the whole roughness profile in terms of both vertical magnitudes and spatial distribution of peaks and valleys. PSD is the Fourier transform of the autocorrelation function of the entire roughness profile, containing both vertical components and spatial distributions. PSD is discussed in detail elsewhere in this book.

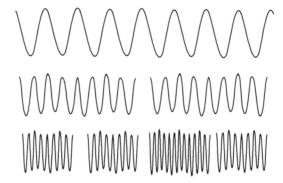

FIGURE 8.12
Surfaces with the same *Ra*, but with different roughness profiles.

8.9 Tolerance

By definition, *tolerance* is the value of the permissible error of a given parametre of the system. In DTM-generated components, dimensional and surface texture parametres (waviness and surface roughness) and their tolerances are considered for measurement and discussion. These tolerances are typically small values, as DTM is the final machining process deployed. The tolerances to be maintained are dictated by the application of the component. Here it may also be noted that waviness and roughness depend on the machining process, work-piece material, tool geometry and its dynamic change as well. Waviness tolerances are often expressed as fractions of mean visible wavelength (λ: 0.633 μm): peak-to-valley (*Pt*) or root-mean-squared (rms) deviation from the prescribed profile locus. Presence of roughness on a DTM surface will result in scatter, which needs to be reduced as per application. To fabricate the diamond turned component within a given set of tight tolerances may be difficult, but not unachievable. However, this comes at the cost of expensive tooling and high production costs. A judicious conclusion is necessary regarding, whether it is justified to spend precious resources to maintain these tight tolerances, and if these tight tolerances are really needed.

It may be of interest to note that the waviness tolerances on injection moulded components (moulded from DTM-generated precision inserts) used for non-imaging applications are generous, often in microns of waviness. The waviness tolerances of MIL grade DTM components (for imaging applications) are generally tight ($\lambda/3 - \lambda/4$). Similarly, consumer optics may have generous roughness tolerances (*Ra* being often a few tens of nm), while DTM components used in space applications need to have low roughness (*Ra* < 10 nm).

8.10 Metrology by Stylus-Based Profilometres

The testing of diamond turned surfaces includes the analysis of *form, figure* and *finish* (roughness) errors. The form and figure errors are often collectively termed surface deviations (or shape errors). The most commonly accepted single-value parametres to evaluate surface deviations and surface finish are peak-to-valley value (*P-V*) and root-mean-squares value (*rms*), respectively. For the surface deviations, more sophisticated parametres like polynomial coefficients have to be used additionally to characterize the deviations. Characterization of aspheric surfaces can be divided into two tasks: surface roughness measurement and the measurement of surface deviations. There are a assortment of roughness measurement techniques available such as total scattering, angle resolved scattering, mechanical and optical profiling, atomic force microscopy, white light interferometry and confocal laser scanning microscopy. Every method has its own advantages and limitations. A nanoscale surface profiler is the most commonly used instrument to characterize the surface roughness, while an interferometer is used to measure the surface deviations in the whole area of the precision surface.

The surface roughness is generally measured by profilometres; mechanical as well as the optical variety. Mechanical profilers are based on a touch probe method, where a stylus makes physical contact with the surface under test. The optical profilers use an optical beam to probe the surface and hence are non-contact by nature. Both probes, mechanical and optical, have some advantages and disadvantages depending on the situation of their use.

The stylus-based profilers have long been in use for measuring statistical properties of smooth optical surfaces (Figure 8.6). They rely on a small-diameter stylus moving along a surface either by movement of the stylus or movement of the surface under test. They are widely used for measuring MEMS, semiconductor devices, optical thin films, optical surfaces and surface finish. A general profilometer consists of a stylus with a small tip, a gauge or transducer, a traverse datum and a processor. As the stylus moves up and down along the surface, it records the heights and valleys of the surface encountered along the scan path, the transducer converts this movement into a signal, and is fed to a processor, which converts this into a visual profile with roughness numbers.

These touch-probe profilers have the disadvantage that the tip of the stylus traverses on the specimen's surface and leaves a mark on it; hence, they are destructive by nature. Moreover, they offer a linear scan of the data, allowing only rotationally invariant error to be detected. Another limitation with the contact profilers is that the contact techniques generally smoothen the surface data. This reduces the bandwidth of the resulting data. Additionally, stylus methods can't test for index inhomogeneities of glass or assembly errors of a lens system.

8.11 Sources of Errors in Surface Quality

The diamond turning process is identified by minimum machining forces (on the work-piece), optimised interaction between diamond tools and work-piece and minimal material removal. Hence, the macro-machining conditions do not have much role to play to decide the surface quality of the diamond turned surface. Consequently, the micro-machining process has some typical sources of surface deviations.

The surface quality of the diamond turned surfaces depends primarily on four factors: (1) work-piece material properties; (2) DTM equipment and machining conditions; (3) tooling parametres; and (4) machining and metrology protocols. The first three issues are already discussed in this book. We deliberate briefly in this chapter, on the surface errors emanating (Figure 8.13) from the protocol violations at various stages of precision component development by DTM, viz: tool setting, work-piece holding and handling, machining and during metrology.

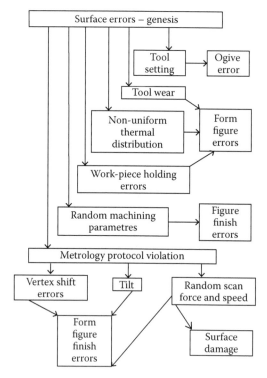

FIGURE 8.13
Surface errors – genesis.

8.12 Ogive Error

In diamond turning operations, tool-setting is one of the most significant stages of precision surface generation. It is necessary to align the centres of the precision tool, work-piece, fixture and vacuum chuck within sub-micron alignment error. If it is wrongly mounted, and if the cutting point of the diamond tool is not aligned with the centre of the work-piece, a variety of errors will manifest on the surface generated, in terms of mis-shape of the component generated. The diamond tool can be misaligned either in the vertical direction (Y-axis) or in the horizontal direction (X-axis) or in both directions with respect to the center of the vacuum chuck.

If the diamond tool is misaligned in the vertical axis, its centre could be either above or below the chuck's centre. If the tool centre is set below the chuck's centre, the centre of the work-piece is undercut and a cylinder (with diameter twice the gap between the centres of the diamond tool and of the vacuum chuck) appears at the centre of the work-piece. If the tool centre is set above the chuck's centre, the work-piece is over-cut and a cone (with diameter twice the gap between the centres of the diamond tool and of the vacuum chuck) will form at the work-piece's centre.

In case the diamond tool is misaligned in the horizontal axis, called *x*-offset, the precision surface post-machining will exhibit a conic error known as *ogive error* [64]. Similar to the height misalignment of the tool centre with respect to the vacuum chuck's centre, the ogive error also occurs in either under-cut mode or in over-cut mode. Ogive error is one of the major shape errors, one should be well acquainted with, and which needs good understanding for its minimisation.

An ogive is the roundly tapered end of a two-dimensional or three-dimensional object. When a circularly arc-shaped surface is being generated around a rotational axis, the ogive error may occur, with its magnitude depending on the amount of offset between the generating curve vetex and the axis of symmetry. Figure 8.14a shows a surface with no ogive error. When the tool centre does not reach the chuck's centre in the *x*-direction, this tool *x*-offset will result in an *under-cut ogive error*. This is shown as a gothic dome (Figure 8.14b) at the centre of the work-piece. Alternately, when the tool passes the chuck's centre in the *x*-direction, this tool *x*-offset will result in a *over-cut ogive error*. This is shown as a double-gothic artifact at the centre of the work-piece (Figure 8.14c).

The ogive error can occur despite (1) a near-perfect coordinated motion of tool and work-piece, (2) near-parallelism of the rotating work-piece and circular arc being generated, (3) zero tool wear and (4) zero machining effects on the surface under generation. In either cases of *ogive error* (both under-cut and over-cut), the artifacts at the work-piece's centre will damage the tool and will shorten its useful life. Also, the *ogive error* affects the tolerances

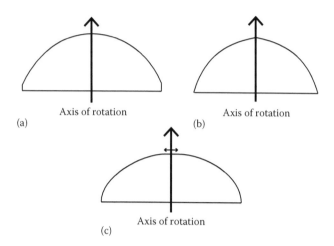

FIGURE 8.14

(a) Surface without ogive error. (b) Surface with under-cut ogive error. (c) Surface with over-cut ogive error.

TABLE 8.1

M-Shapes and W-Shapes at the Work-Piece's Centre Due to Ogive Error

Profile	Ogive Error at the Work-Piece's Centre	
	Tool Stopped Before Spin Axis	Tool Past Spin Axis
Convex	M-shape	W-shape
Concave	W-shape	M-shape

of the precision components developed by diamond turning operations, by directly impacting the waviness error. Hence, it is necessary to assess the ogive error while tool is being set and necessary steps need to be taken to reduce this error to a possible minimum magnitude, in order to meet the tolerance criteria of the component. The diamond tool x-offset will result in opposite ogive errors in convex and concave profiles, manifested as W-shape waviness and M-shape waviness at the profile centre (Table 8.1).

8.13 Metrology Errors

More often than not, the diamond turning offers a precision-machined surface, provided all the necessary precautions are taken before and during machining [38]. However, many errors may creep-in during the handling of the work-piece, which may lead to a damaged component [65]. Hence, care may be taken to handle the precision diamond turned component properly. Another major source of surface errors is during the surface characterization stage itself. There may be many situations when the diamond turning is done in near-ideal conditions with the best machining parametric matrix with a diamond tool with practically no wear, and still the surface may show unacceptable surface quality in terms of surface waviness [66]. The culprit may not be the machining but the conditions of surface characterization. Let us consider the process of a precision machined work-piece being characterized by a contact profilometer. This involves: (1) the removal of the work-piece from the vacuum chuck (along with/without its fixture)/jig; (2) placement of the work-piece, with/without its fixture beneath the stylus of the profilometer; (3) setting of the default force of the profilometer's stylus to scan the surface; (4) selection of the vertex point (highest/lowest as per the surface curvature profile) of the precision machined surface by exploring and cresting; (5) design data inputs into the metrology code to generate the target surface; (6) selection of suitable scan length, evaluation length and cut-off length for the surface analysis and (7) performing a linear scan on the surface and (8) analysis of the scanned data. This means that, at each step, one may commit an unintentional mistake and may end up with either a flawed surface or misrepresented surface profile error, which may render the surface unacceptable by the user (while, in reality, the surface may have an acceptable profile error).

When the component is released from the vacuum chuck along with/ without its fixture, the release of the pressure on the surface may alter its profile significantly, leading to form error, in terms of its radius of curvature and conic constant. Figure 8.15 explains the holding error scenario. It is very interesting to see the extents of the holding errors for different materials during DTM operations. It may also be noted that these holding errors will significantly impact the profile error metrology as well as the profile error compensation. This situation can be well avoided by using a suitable fixture during diamond turning, where the vacuum pressure doesn't change the form of the work-piece.

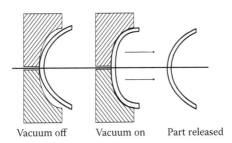

Vacuum off   Vacuum on   Part released

FIGURE 8.15
Holding errors during DTM.

The next crucial step of DTM surface characterization involves its positioning under the profilometer (with/without its fixture) for scanning. If the base of the component/fixture is skewed, this will result in a tilt in the surface to be scanned, as shown in Figure 8.16. Some of the modern post-scanning analysis software have a tilt removal option. This option may be judiciously used, by investigating whether the tilt occurred during the fabrication stage or due to its erroneous placement at the profilometer. If the component is non-parallel or trapezoidal in shape (having some wedge), it needs to be addressed (to make it parallel) by machining off the wedge of the work-piece/fixture with respect to the vacuum chuck prior to the diamond turning of the (desired) surface profile. However, this may result in reduction of the thickness of the work-piece. Therefore, adequate thickness allowance may be designated on the blank of the work-piece.

During the scan of the work-piece by the stylus, any of these three scenarios may emerge: (1) The stylus's force (with which it is placed on the work-piece) is improper, (2) the speed (at which it moves across the work-piece) is unsuitable or (3) both (stylus force and speed of scan) are not properly selected. The result is: the surface may either get damaged by scratching (over-force and drag) or the stylus may skip (under-force and jumping) over the peaks and valleys of the profile. In either case, one ends with a rejected component/wrong result.

Another type of error that can occur at the metrology stage is the positioning (of the work-piece under the profilometer) error, resulting in a lateral shift of the vertex of the component. This will cause a grave situation with (1) an unequal clear aperture measurement and (2) misregistration of the fabricated conic profile and its matching with the targeted conic profile. Thus, the conic (as per scan) presented for analysis will have a different base RoC and a different conic constant compared to the conic which is fabricated and intended to be analysed. This will naturally lead to an unacceptable profile error. Figures 8.16 and 8.17 present these anomalies.

The lateral placement of the work-piece may not affect a flat machined component in terms of its waviness but will certainly dent the analysis of its useful aperture. In case of conics and spherical components, the shift of

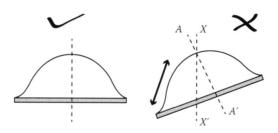

FIGURE 8.16
DTM components placed under the profilometer with a skewed base.

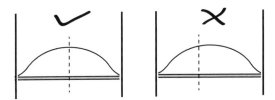

FIGURE 8.17
Profile change due to vertex shift during profilometry.

aperture and vertex will end up in large magnitudes of waviness errors. These errors can be avoided by proper cresting of the vertex of the component under the profilometer.

8.14 Thermal Effects and Metrology

During machining, the work-piece and the diamond tool will interact, the tool shears through the material of the work-piece and material is cleaved. This will surely give rise to a significant amount of thermal energy. The use of coolant in diamond turning will certainly take the heat from the tool-job interface. However, the heat that has crept into the work-piece may remain there for some time until it gets dissipated, depending on the thermal diffusivity of the material of the work-piece. With repeated cycles of (rough and fine) machining, the temperature at the tool–job interface rises, the heat gets trapped in the work-piece with the residual heat (for the next cycle) forming a Gaussian thermal profile. This causes bulging of the material and de-shaping of the surface profile. Hence, a process needs to be established to compensate for the deterioration in the surface profile (due to the heat thus generated) and thereby to modify the tool path accordingly. This process may include (1) calculation of heat generated in the work-piece, (2) thermal modelling in steady state, (3) generalisation of the model to the transient state, (4) validation of the model by measuring the temperature at the tool–job interface, (5) computation of the change in the surface profile due to the thermal accumulation in the work-piece and finally (6) modifying the tool path in the final machining cycle to account for this profile variation. This exercise varies from internal zone to external zone (through the middle portions) of the work-piece material, from material to material as per their thermal properties (diffusivity and expansion) and as per the prescribed tolerances of the components under machining.

The thermal flux thus generated, part-transferred and part-trapped in the work-piece, results in the swelling of the work-piece. This swelling is often followed by hysteresis losses after recovery, resulting in deterioration of the surface profile, both in terms of surface waviness and roughness.

8.15 Error Compensation Techniques

The surface quality improvement depends primarily on the design requirements and on the correction process formulated and adopted in the development of the precision components. While the finally achievable roughness depends on the machine dynamics, tool geometry and on its imprints on the work-piece, waviness reduction is a major concern. While a great amount of waviness correction takes place by (1) judicious selection of the right diamond tool and monitoring its wear dynamics, (2) machining with optimal combination of machining parametres, (3) control of the machining process and (4) adherence to prescribed machining-metrology protocols, still a significant amount of waviness remains on the work-piece. Depending on the stringency of the surface quality, this waviness needs to be further reduced so as to adhere to the prescribed tolerance limits. This calls for a tool path compensatory approach, wherein the tool path is slightly adjusted as per the waviness present on the surface to deliver reduced waviness.

In this tool path compensatory approach, the metrology initiates the compensation process. The profile of the work-piece is a result of its sagitta (depth/height) variation with respect to its aperture (given in the sag table at the diamond turning stage). Waviness or the profile error of the diamond surface as defined earlier is the mismatch of the designed and fabricated profiles. In other words, the waviness is the deviation of sagitta (as diamond turned) from the prescribed sagitta (as per design) with respect to the aperture. In tool path compensation, this sag deviation versus aperture is fed back (Figure 8.18) to the prescribed tool path, thereby (slightly) changing the tool path [67]. The new tool path may be slightly different from the original profile equation. However, it has the negative waviness component included, and is expected to deliver a near-zero waviness surface in its next machining

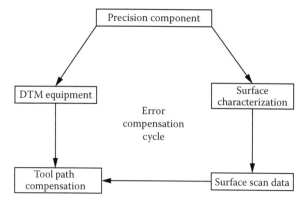

FIGURE 8.18
Schematic of profile error compensation.

cycle (presuming that tool dynamics do not change significantly in successive machining cycles). Zero error may not be achievable within one compensatory cycle. But within two to three machining cycles, the waviness can be reduced to the tolerable limits.

It may be noted that this approach works only for *figure error* and may not work for *form error*, where the tool geometry may be a major culprit. Form error correction needs a different approach (due to tool geometry dynamics during successive machining cycles). Waviness reduction depends primarily on the accurate characterization of the diamond turned work-piece surface. Hence, it is incumbent on the metrology process to deliver a truthful surface profile, devoid of the errors that may creep in during the surface characterization stage. This calls for a careful handling of the work-piece post-diamond turning followed by establishment and adherence to a standardised metrology protocol.

The success of precision surface development is based on fabrication (diamond turning in the present case) as much as on the surface characterization methodology. In this context, study of the surface quality is both challenging and rewarding exercise.

In this chapter, an insight of various topics towards qualification of the precision component developed by DTM, in terms of geometrical dimensions and surface features is presented. As per the application (where the precision component is deployed), the degree of precision desired and achieved, and as per the prescribed tolerance ranges of the quality criterion, the DTM practitioners are advised to dig deep in to the domain of surface characterization, and to learn from the on-the-field experiences. After all, the knowledge generated and skills thus acquired will have larger impact in real-life problem-solving.

8.16 Summary

This chapter deals with the qualification of the components fabricated by DTM, as the latter process complements the fabrication stage. The main features of dimensional measurements and the surface characterization, namely, form, figure and finish, are dealt with in this discussion. The precision surface metrology processes are explained in terms of the surface texture parametres with relevant definitions and their correlation with the proposed surface quality criteria. A brief presentation of tolerances in DTM is included to give an idea of accuracy and precision requirements and on how they are achieved by DTM. The concepts of contact profilometer and the corresponding methods of precision surface scanning and analysis are explained. Additionally, the sources of surface errors (apart from fabrication) during precision component development are described. The chapter concludes with a discussion on how to improve the surface quality in terms of its major criteria: form, figure and finish.

9

Advances in DTM Technology

9.1 Introduction

This chapter reviews very recent and on-going interests and developments being reported from the world-over in the area of DTM. Almost all aspects of DTM have drawn the interest of researchers: machine tool structure, cutting tool, work material, process monitoring, tool holder, path planning, effect of coolant etc. Focus on choosing the reports is more on those that deal with applied technological developments rather than fundamental and routine investigative-type research. In addition, the technological developments are explained in a simple manner with some basic schematics to elucidate the concept that is being pursued and is in reports. Figure 9.1 shows the recent areas of research in DTM.

9.2 DTM Process Monitoring

The DTM group in Hong Kong Polytechnic (led by W. B. Lee) has reported [68,69] attempts in developing techniques to monitor surface roughness during the DTM cutting process. Their goal is to effectively correlate force signals to suitable surface profile signals. Surface profiles are quantified in a unique way along the cutting path (spiral), besides taking surface profiles in the conventional manner radially along the feed direction (Figure 9.2). Using a centrally placed hole to trace the start and end of the spiral motion, and using high feeds per revolution in order to accentuate the spirals, the surface profile along the spiral is extracted from the surface profile maps and is correlated with force signals by matching the signals temporally. It is seen that the surface undulations along the spiral path correlate well with the force variations. In a related and perhaps follow-up study, vibration sensors are placed and vibration signals are collected from the DTM process. Again using the method as explained above, it is seen that the vibration signals correlate well with the surface profile undulations along the cutting path.

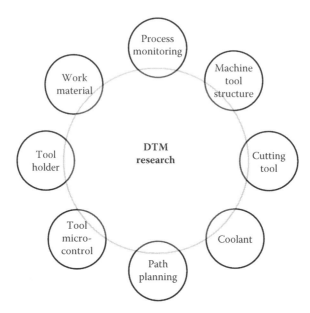

FIGURE 9.1
Areas of recent research in DTM.

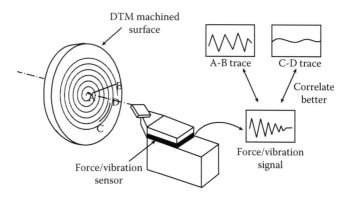

FIGURE 9.2
Surface roughness profile along the cutting velocity path (spiral path) shows better correlation to force and vibration signals.

Hence, by monitoring the vibration or force signals, clues about variations in surface profile (along the cutting spiral path) can be obtained.

A method of monitoring the life of the diamond tool is also proposed [70]. The objective is to measure changes in the nose radius of the tool by performing taper grooving experiments and measuring changes in the groove profile (measured using a non-contact surface profiler). Since elastic recovery is inevitable, the groove profile will not match that of the tool nose radius.

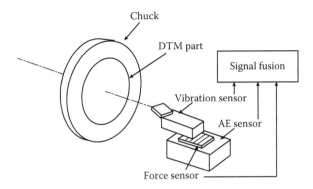

FIGURE 9.3
Use of several sensors in DTM and novel methods of fusing the signals.

An attempt has been made to estimate the elastic recovery, using volume constancy and some simple assumptions on the elastic recovery of the groove profile. Experiments to test this method have yielded reasonable results. However, the usefulness of this method to effectively monitor the tool condition during a DTM process is to be seen.

A sensor fusion approach has been proposed [71] for monitoring the DTM process. A combination of sensors – force, vibration and acoustic emission – is placed as near to the cutting tool as possible (Figure 9.3), and data from all these sensors is collected simultaneously. Regular and simple (linear and Gaussian based) signal processing and statistical methods cannot be used in DTM processing, as even a sample signal in the DTM process is complex in terms of frequency, time and state-space domains. Hence, a new adaptive nonparametric Bayesian Dirichlet process method of fusing these sensor signals is proposed, tested and compared with other fusion methods.

9.3 Developments Related to Machine Tools

A new type of architecture for the DTM machine is being proposed [72], where the cutting tool does not simply move on a linear axis, but is made to swing by a rotary arm (Figure 9.4). This machine thus has two linear axes and three rotational axes. Two of the rotary axes motions are driven by hydraulic motors. The basic principle is that the swinging arm holding the cutting tool can be made to execute an arc motion in planes of any inclination; thereby a spherical surface arc on the rotating work-piece is traced. The orientation of the plane needed for this can be calculated geometrically. It is claimed that interpolation errors normally seen in the T-type structure of the DTM can be

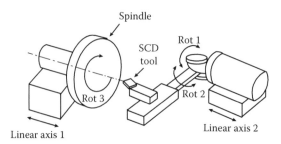

FIGURE 9.4
Structure of swing arm DTM machine.

avoided by using this machine tool architecture. A kinematic error model is developed to undertake sensitivity analysis along an alignment method for the two hydraulic motors to reduce the effect of various errors.

The Fraunhofer Institute has explored [73] an unusual idea of a two-sided DTM machining (one side machines the top surface and other the bottom side) simultaneously, of a work-piece that is uniquely clamped and held in the middle (Figure 9.5). This requires significant modification of the machine structure with two Z-axis slides on either side of a centrally placed spindle. The spindle has to be accessible from two faces – which means the spindle face plate has to be actuated from the bottom end. The motive behind these efforts is to make the DTM process cost-effective. Introduction of automation in the DTM process (e.g. loading work-piece, tool change, etc.) is also being explored.

The issue of imbalance of spindle continues to attract attention and methods are being devised to study and counteract its effects [74]. This work reports

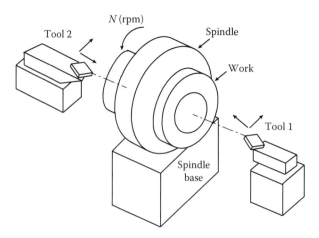

FIGURE 9.5
Unique two-sided parallel machining to reduce DTM process times.

modeling the dynamics of an aerostatic spindle and then using this understanding to counteract the effects. Another group reports [75] that imbalance effects can lead to the formation of a star effect seen on DTM machined surfaces with radial spokes emanating from the center. The formation of such spokes is attributed to what is called as air-hammering effect coupled with the air bearing spindle effects.

In a related development, Brinksmeier's group in University of Bremen is suggesting [76] the use of high speed spindles in the DTM process. The DTM process, in its current configuration, requires accurate balancing of spindles. Such a requirement becomes even more stringent (faster and more precise) at high speeds. Novel methods are being pursued to overcome this problem, while reducing stiffness in spindle mounts, to increase sensitivity of unbalanced measurements.

Efforts also continue to be reported, in modeling errors in DTM machines: both in constituent modules (spindles, slides etc.) and the entire structure as a whole. Such models are exploring the error-compensation and form accuracy improvement.

9.4 Developments Related to Cutting Tools

Another effort [77] focuses on a new form of polycrystalline diamond tool, by aggregating nano-crystalline diamond particles without any binder (Figure 9.6). Such an aggregate is shown to be denser and 11% less

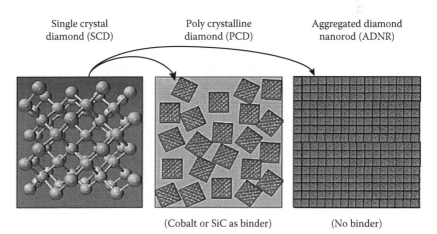

FIGURE 9.6
Aggregated diamond nanorod (ADNR) – a new material for cutting tools harder than single crystal diamond.

compressible than single crystal diamond (SCD). This new material, called ADNR, has better fracture toughness, Knoop hardness and wear coefficient, but lower Young's modulus, than SCD. Tools made of a similar nano-crystalline form of diamond are being offered [78], claiming at par performance with a SCD-based cutting tool. However, these claims need to be verified.

It is well known that a new cutting tool performs well in DTM machining of brittle materials, but the performance degrades with time. This is often attributed to tool wear. Recent studies [79] investigate on how tool wear affects the ductile-to-brittle transitions in DTM machining of brittle materials. Using a simulation model that uses experimental tool wear contours, it is shown that the hydrostatic stress distribution – that helps in avoiding crack propagation – is disturbed as flank wear progresses; particularly the hydrostatic stress fluctuates as the tool wears out. Besides, the highest stress point also is shown to be displaced behind the tool edge. Such an analysis can show pointers on how to perhaps continue to maintain ductile conditions even when the tool wears.

Another effort by the Brinksmeier-led group at University of Bremen reports [80] a novel thermal actuator to move and align diamond cutting tool edges (Figure 9.7). The motivation for this work is the use of multiple cutting edges in the DTM milling process. Milling requires precise alignment of various cutting edges. Since depths of cut are very small in DTM, they are all usefully participating in surface generation. DTM milling with several cutting edges can improve process cycle times. Using a flexure arrangement for the milling tool with multiple diamond tools, LED-based infrared heating is utilised to radially move the tool edges from a few nanometres to hundreds of nanometres. The testing of this type of cutting tool is reported to be in progress.

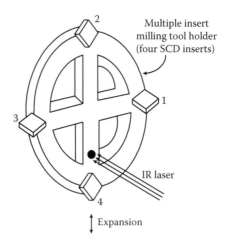

FIGURE 9.7
A unique idea of thermally actuating various inserts in DTM milling process to align the inserts perfectly.

W. B. Lee's group in Hong Kong reports [81] a method to optimally choose the best cutting tool suited for a given optical surface to be generated via the DTM process. It is known that the cutting tool nose radius influences the form that is generated. While having a zero nose radius would be best, it would be practically not feasible; in practice, one would like to have a large possible nose radius to maintain tool strength. In order to balance these two conflicting requirements, the group uses ray-tracing models to predict the optical errors that can result from the use of various tool nose radii. The form is first predicted for a given nose radius and then fed into the ray-tracing model. An optimisation search is then initiated to maximise the nose radius which may provide an acceptable optical error (e.g. wave front aberration).

9.5 Influence of Coolant in DTM

Several groups are testing the effects of various types of coolants and additives on the performance in the DTM process. Some key findings are reported in this section.

Fritz Klocke's group at Fraunhofer reports [82] that the critical depth of cut, where ductile-to-brittle transition occurs, can be extended by optimising the coolant used in diamond turning. Taper scratch tests on a hard brittle material (e.g. tungsten carbide – WC) are performed in the presence of various cutting fluids. It is seen that, under some specific cutting fluids, the critical depth of cut is higher. It is advised that customised cutting fluids are to be used to enhance efficiency of DTM processing. It is also reported [83] that pH levels of the water used in the coolant affect the critical depth of cut, based on tests on WC and two glass materials.

W. B. Lee's group reports [84,85] the use of nano-droplet-enriched cutting fluids (NDCF) in DTM and in taper scratch experiments using diamond tools on 6061 aluminium alloy. NDCF is a mix of mineral oil and water with the oil in the form of nano-droplets (Figure 9.8). In DTM experiments conducted with NDCF, surface finish is seen to be improved. In the taper scratch (2 micrometres to 0, 0.01° taper), it is reported that, under the influence of NDCF, the groove is uniform with no plastic deformation bumps observed. Such bumps are, however, observed under conventional cutting fluids. It is argued that NDCF has better thinning capability and hence can enter the tool-chip interface, reducing friction and enhancing chip formation, and material deformation near the cutting edge.

As mentioned elsewhere in this book, the problem of inability of diamond tools to effectively machine ferrous alloys continues to be a challenge; it seems to constantly motivate researchers for a solution. Recent work by Rensselaer Polytechnic reports [86] the use of graphene oxide platelets in a semi-synthetic cutting fluid, while machining ferrous alloys with PCD tools.

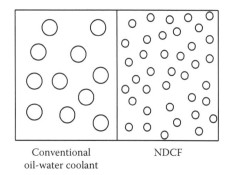

Conventional   NDCF
oil-water coolant

FIGURE 9.8
Use of nano-droplet enriched cutting fluids in DTM has been reported.

Though this is not tested in the DTM process, mild steel is machined with a PCD tool. The results have shown a 30% decrease in tool wear, 50% reduction in cutting temperatures and 30% decrease in cutting forces. Based on X-ray photoelectron spectroscopy (XPS) studies, it is felt that the graphene oxide particles are interfering with the carbon diffusion normally seen at the tool-chip interface. This technique could be extended and applied to the DTM process also.

9.6 Vibration-Based Controlled-Tool Motion

Efforts are in progress to extend the range of the fast tool servo (FTS) system. In one such attempt [87], slow-tool servo (STS) and FTS are synchronised so as to increase the reach of the tool. In this system, the C-axis is synchronised with the Z-axis slide and with the FTS system. The STS will be used to 'macro' position the cutting tool, while the FTS will be used locally at that position to execute necessary 'micro' small-range cutting motions. The reports indicate changes in programming techniques, to smartly incorporate the use of both STS and FTS motion paths. Several examples of increased versatility in geometrical shapes are also reported. It is, however, noted that dynamic performance of the system is still limited by the STS.

Meanwhile, efforts are continuing to extend the motion stroke of the FTS system. A novel long range FTS system (single axis) based on voice coil motors is reported [88]. The motors drive a linear air bearing stage with a closed loop control using glass linear encoders. The mechanical design is optimised to have first and second modal frequencies of above 1 kHz. This system is reported to have a stroke of up to 30 mm, with accelerations approaching 100 Gs. It can be used up to a frequency of 200 Hz.

Single axis FTS systems are giving way to 2-axis (and possibly even 3-axis) FTS systems. A 2-axis FTS is reported [89] for machining dimple patterns on rollers used in large-area hot embossing. One axis provides in-feed (perpendicular to roller axis) of about 40 micrometres, while the other is parallel to the roller axis with a stroke of 75 micrometres. The system is designed for an operational bandwidth of 1.5 kHz.

In an interesting report [90] of application of a 2-axis FTS, an attempt is made to undo the characteristic deterministic cutting motion of the DTM process. Using a flexure-based FTS with decoupled motions in two axes of motion, pseudo-random vibrations are introduced to disrupt the characteristic spiral feed marks on the surface of the work-piece. This technique is implemented by developing a new surface topography generation algorithm. The resulting tool paths are experimentally tested and compared without the pseudo-random vibration. Light scattering tests are also performed to show that application of random vibrations suppress the scattering fringes normally seen on DTM machined surfaces.

In a deviation from using induced-vibration to control and synchronise tool motion, vibrations can also be used to cause local loss of tool-chip interface to reduce wear and forces. This technique has been popularly developed and applied much earlier to reduce tool wear, while using diamond tools to diamond turn ferrous materials. Commercial success of this technique has however been limited with one known company (e.g. Delta Optics, Singapore) utilising this technology effectively. This technique has been recently demonstrated with some reported success in diamond turning of titanium alloy [91].

Vibrations to cause tool-chip intermittent contact losses are largely based 'elliptical vibration cutting' where, the tool makes an elliptical cutting motion at high frequencies. In one small part of the elliptical arc, the tool is in cut with the rest of the arc path out of cut. Such elliptical vibration motions are often achieved using resonant elastic-deformation-based mechanical vibration systems triggered by piezo-actuators. Resonant systems have inherent limitations of cross-talk between axes of motion, limitations in frequency adjustments and being sensitive to manufacturing and material property differences from design plans. In order to overcome this problem, non-resonant systems are designed. Such a non-resonant system can execute 3D motion of the cutting tool with four piezo-stacks is tested in the DTM process [92]. Being non-resonant in nature, the signals to the piezo-stacks can be varied to adjust amplitude frequency, phase shifts and acting locations (Figure 9.9). In another development, the elliptical vibration cut incorporated in a DTM machine is combined with an STS system and is reported as a double-frequency vibration cutting [93] (see Figure 9.9). It is argued that, because of the resulting unique motion of the tool, the cutting plane does not always cross the work-piece center. To facilitate this motion, a new algorithm of tool path generation, different tool geometry selection and surface prediction techniques is proposed.

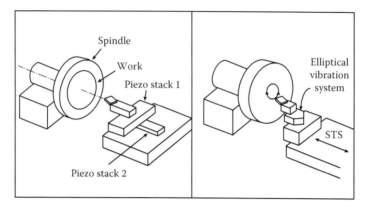

FIGURE 9.9
Elliptical vibration machining has been adopted in DTM in various ways.

To conclude this discussion on machine tool vibrations and reviewing the work done, it is reported [94] that active vibrations tactfully induced in the cutting tool may cancel the effect of other vibrations in the machine tool and process system.

9.7 Tool-Path Planning

New tool-path planning techniques as reported are inspired by novel geometries that are required to be diamond turned. Besides, the challenges to reduce both process-planning and processing cost have also pushed the boundaries for novel algorithms for tool path generation.

In an interesting exercise [95], CAD/CAM software used in conventional machining is deployed for tool path planning in DTM processes. The DTM-specific CAM software for the DTM tool path planning comes at a high cost. To overcome this, an application programming interface (API) is developed. This API works with the commercial CAD software (SolidWorks). Conjoining these together, spiral tool path trajectories are developed to generate path plans for free-form surfaces using FTS/STS systems. This interface has been effectively deployed to generate tool paths and thereby to fabricate the components with free-form profiles and micro-optic arrays using STS/FTS systems.

Optical components often require arrays of identical micro-features which repeat discontinuously. A new tool path planning algorithm is developed [96], which deviates from the typical spiral path planning routinely employed for FTS/STS-based feature generation. Using this algorithm, a virtual spindle axis is created at each feature of the array. Then the tool planning is carried

out as if only the STS/FTS system machines that feature alone. In this complicated array, a few micro-optics arrays (MOAs) are machined with greater
ease of planning the tool paths.

A new spiral tool path generation method is reported [97]. Instead of projecting a planar Archimedean spiral on the work surface, which leads to the
problem of unequal spiral spacing, a surface of revolution close to the surface
of the work-piece is first generated. An Archimedean spiral is then generated
on this surface of revolution with equal spacing. This surface spiral is then
projected along its normal surface direction on the desired work-piece surface.
This way, the spacing variation is reduced considerably. Detailed algorithms
for automating the surface of revolution, spiral generation and projection are
developed and tested on actual work surfaces with encouraging results.

A novel tool motion planning approach is proposed [98] to machine a Fresnel
lens on a cylindrical roller. In a normal FTS arrangement in a DTM machine,
the C-axis and the W-axis (Z-axis of the FTS stage) are synchronised. Also, it is
usually arranged such that, the X-axis is also synchronised leading to an at-best
3-axis synchronisation arrangement. However, such an arrangement is not
sufficient to create the required Fresnel lens on the surface of the roller. To overcome this problem, the machine is modified to synchronise even the B-axis of
the lathe with the C-axis leading to synchronisation of all four axes (Figure 9.10).
This requires a novel tool path planning method. This tool path is successfully
implemented to develop the much-needed Fresnel lens on the roller.

A different method, while still utilising the B-axis, is proposed [99] indepen-
dently, to process regular Fresnel lenses on flat surfaces. The circular swing of
the B-axis and a custom curved tool edge are used to suit the Fresnel and roller
geometries, to generate the Fresnel lens geometry.

Tool path planning algorithms have evolved over time, to optimally use
the hardware developments such as 2-axis and 3-axis FTS systems. In a

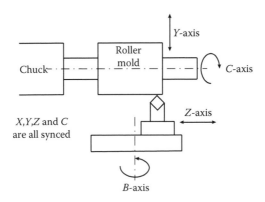

FIGURE 9.10
Unique methods to make complex shapes such as Fresnel array on the cylindrical surface of
the roller mold.

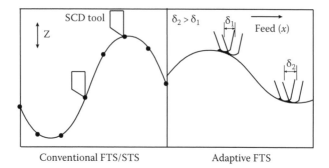

FIGURE 9.11
Some unique uses of the FTS system to optimise tool paths and reduce process times.

normal DTM process, the feed motion is along the X-axis; the resulting Archimedes spiral motion is equally spaced based on the feed per revolution or on the angular rotation. The question that is addressed here is, whether optimising the spiral gap can increase the efficiency of cutting. Another development [43] uses the 2-axis FTS to improve the efficiency of path planning (Adaptive Tool Servo, ATS). In a 2-axis FTS, the second axis is normally along the X-axis, used to feed the cutting tool. Since micro-motion control along this axis is now available, based on the curvature of the work-piece, it is proposed to locally alter the feed. This facilitates the manipulation of the feed as per the quantum of curvature changes: at small curvature change locations, the feed can be increased. At rapid curvature changes, the feed is decreased (Figure 9.11). Detailed algorithms are developed to implement and test this method for various work geometries. Significant reduction in the number of cutting points, machining time and form errors are demonstrated.

Efficient tool path planning can also involve analyzing various tool path planning methods and choose the best one, *a priori*. This has been suggested by Rahman's group at National University of Singapore [100]. They advocate a profile error analysis (PEA) where the two strategies of constant angle and constant arc are used to analyze the desired surface profile to be generated. Based on this, the method that gives the least number of cutting points to meet the form requirements is then chosen.

9.8 New Materials and Materials Treatment

As new families of materials are being diamond turned, efforts are in progress in terms of material property alterations, prior/during the process, to suit the DTM process.

New grades of aluminum alloys suitable for optical applications are evolving. One example is the development of rapidly solidified cast aluminum alloys that have very refined microstructure. DTM-based machinability studies are in progress for such materials. In a collaborative effort between researchers in South Africa and Taiwan [101], machinability experiments are conducted on alloy RSA 905 optical grade aluminum. The study reports that roughness is decreased with cut length and the lowest achievable surface roughness is a few nanometres over a 4-km cutting length.

Potassium dihydrogen phosphate crystals are reported [102] to be processable by the DTM process. This material easily absorbs water, has very low fracture toughness and is highly sensitive to thermal changes. In a collaborative effort between China and the United Kingdom, a manufacturing sequence is developed, starting with the DTM process, followed by polishing and then ion-beam figuring to provide needed shape.

DTM machinability studies on a Cu-Cr-Zr alloy are reported [103]. This hard copper alloy, used in molds, dies and optical components, contains hard precipitates and surface oxides, which make it challenging to machine. Chip morphology and tool wear studies are conducted; acoustic emission signals (from the process) are analysed; abrasive and notch wear on the diamond tool are also studied.

Special treatment of materials to improve their machinability has been a topic of interest in general in conventional machining. This has been extended specifically to the DTM process also. The efficacy of electro-pulse treatment (EPT) of titanium alloys on the ability to DTM machine these materials is explored [104]. Titanium alloys such as Ti-6Al-4V, while having great commercial interest, are notoriously difficult to machine both conventionally and more so using the DTM process. The EPT system is customised and titanium-alloy samples are subjected to EPT under various conditions (mainly frequency). EPT treatment is reported to reduce grain size and to soften the material. Then the samples are diamond turned, the effect of EPT on chip morphology and the machined surface condition are studied. It is reported that EPT can substantially improve the chip formation conditions and hence improve both surface finish and associated tool life.

In-situ (i.e. during machining) treatment of materials to facilitate DTM has also been reported. The treatment reported is that of heating the material to improve its ductility by a laser beam applied directly to the cutting edge. The first reported work is that of John Patten's group at Michigan State University [105]. In this method, an IR laser beam is made to travel with a designed attachment, through the diamond cutting tool and made to reach the edge directly, where it is focused and the material closest to the cutting edge is heated-up. This method is tested on many materials, improvements in material removal rates and surface finish/integrity are observed. Based on this development, the DTM machine company Precitech has also recently showcased a commercial add-on that allows users to use laser assist in the DTM process.

In an independent development, W. B. Lee's group also has reported [106] the use of a laser beam (100 micrometres in diameter) to treat material ahead of the cutting edge, by a special microscope arrangement.

9.9 Tool Holding for DTM

A novel idea of using a hexapod style parallel kinematic system to hold and move the cutting tool (Figure 9.12) on a DTM machine has been recently reported [69]. The aim here is to improve the accuracy of older DTM machines and compensating for the errors. In addition, aspects such as tool-normal machining can be achieved using such a system. Since the kinematic motion is different in the hexapod, new algorithms for tool path generation are developed for this system. Experiments involving simple face turning and complex patterns such as that seen in a compound eye are used to demonstrate the efficacy and viability of such a system.

The fairly popular DTM machine builder, AMETEK Precitech, along with a German company, Levicron Gmbh, have recently reported efforts to improve tool holding and work holding in a DTM machining setup with the intention of moving towards automation of tool and work changes. They report automatic taper clamping systems with 0.2 micrometre repeatability, static run-outs of 0.6 micrometre or less and sufficient balancing requirements at spindle speeds of up to 60,000 rpm.

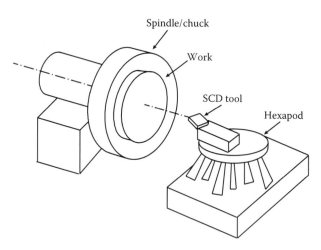

FIGURE 9.12
Proposed concept of using a hexapod to control the cutting tool position.

9.10 Summary

As the reader can see from the discussion in this chapter, novel developments are taking place globally in various aspects of the DTM process. In Europe, there seems to be a conscious effort to improve the economics of the DTM process by reducing processing times. Radical changes are being proposed towards this goal. Techniques from conventional machining are being adopted and customised for DTM processing. Novel ideas to improve the DTM process are also being driven by specific machining requirements. This chapter presents a bird's view of the work done globally. Interesting studies on coolant enhancements have also been attempted in the DTM process. It is seen that in the near future, the DTM process will undergo significant changes to make it more mainstream for adoption in various applications. As in all spheres of technology, it is necessary to have vision, think innovatively and test these novel ideas.

9.11 Questions

1. Besides the developments listed in this chapter, think through and list five possible new directions of developments and research possible in DTM.
2. List five recent developments in conventional machining automation. Can you extend them directly to DTM? If not, how would you customise these developments to make them applicable to DTM?

Bibliography

1. Venkatesh, V.C. and S. Izman. 2007. *Precision Engineering*. New Delhi: Tata McGraw Hill.
2. Taniguchi, N. 1994. The state of the art of nanotechnology for processing of ultra-precision and ultra-fine products. *Journal of the American Society of Precision Engineering* 16(1): 5–24.
3. Shore, P. and P. Morantz. 2012. Ultra-precision: Enabling our future. *Philosophical Transactions of the Royal Society of London A* 370: 3993–4014. http://rsta.royal societypublishing.org/content/370/1973/3993 (accessed January 1, 2017).
4. Xiao, X. 1989. Ultra high precision machining techniques – Applications and current status. Air Force Systems Command Report. http://www.dtic.mil/dtic /tr/fulltext/u2/a233532.pdf (accessed January 1, 2017).
5. Gerchman, M.C. 1986. Specification and manufacturing considerations of diamond machined optical components. *Proceedings of SPIE* 607.
6. Lee, W.B. and B.C.F. Cheung. 2003. *Surface Generation in Ultra-Precision Diamond Turning: Modelling and Practices*. London: Professional Engineering Publishing Limited.
7. Chapman, G. 2001. Ultra-precision machining systems: An enabling technology for perfect surfaces, Moore Nanotechnology Systems LLC. http://nano technology.com/technology (accessed January 1, 2017).
8. Empire Precision. 2014. SPDT eBook. http://www.empireprecision.com/blog /topic/single-point-diamond-turning-spdt.
9. Rhorer, R.L. and C.J. Evans. 2009. Fabrication of optics by diamond turning. *In hand book of optics*, ed. Michael Bass, Vol. II, Part 2, Chapter 10, 3rd edition. New York: McGraw-Hill.
10. Jain, V.K., A. Sidpara, M. Ravisankar, and M. Das. 2016. Micro-manufacturing: An introduction. In *Introduction to Micromachining*, 2nd ed., V.K. Jain, Ed. New Delhi: Narosa Publishing House.
11. Kumar, J., V.S. Negi, K.D. Chattopadhyay, R.V. Sarepaka, and R.K. Sinha. 2017. Thermal effects in single point diamond turning: Analysis, modeling and experimental study. *Measurement* 102: 96–105.
12. Walter, M., B. Norlund, R. Koning, and J. Roblee. Precitech, Inc. Keene, NH 03431, 2014. Error Budget as a Design Tool for Ultra-Precision Diamond Turning Machines Form Errors. http://www.precitech.com/downloads/Error Budget as a Design Tool For Ultra-Precision Diamond Turning Machines Form Errors.pdf (accessed November 10, 2016).
13. Takasu, S., M. Masuda, T. Nishiguchi and A. Kobayashi. 1985. Influence of study vibration with small amplitude upon surface roughness in diamond machining. *CIRP Annals-manufacturing Technology* 34(1): 463–467.
14. Lee, W.B. and C.F. Cheung. 2001. A dynamic surface topography model for the prediction of nano-surface generation in ultra-precision machining. *International Journal of Mechanical Sciences* 43: 961–991.

15. Balasubramaniam, R. and V.K. Suri. 2011. Diamond turn machining. In *Introduction to Micromachining*, V.K. Jain, Ed. New Delhi: Narosa Publishing House.

16. Patterson, S.R. and E.B. Magreb. 1985. Design and testing of a fast tool servo for diamond turning. *Precision Engineering* 7(3): 123–128.

17. Rahman, M.A., M.R. Amrun, M. Rahman and A.S. Kumar. 2016. Variation of surface generation mechanisms in ultra-precision machining due to relative tool sharpness (RTS) and material properties. *International Journal of Machine Tools and Manufacture* November: http://dx.doi.org/10.1016/j.ijmachtools.2016.11.003.

18. Taniguchi, N. 1996. *Nanotechnology*. Oxford: Oxford University Press.

19. Liu, X., R.E. DeVor, S.G. Kapoor and K.F. Ehmann. 2005. The mechanics of machining at the microscale: Assessment of the current state of the science. *Journal of Manufacturing Science and Engineering* 126(4): 666–678.

20. Nakasuji, T., S. Kodera, S. Hara and H. Matsunaga. 1990. Diamond turning of brittle materials for optical components, *Annals of the CIRP* 39(1): 89–92.

21. Komanduri, R. and L.M. Raff. 2010. Molecular dynamics (MD) simulations of machining at the atomistic scale. In *Introduction to Micromachining*, V.K. Jain, Ed. New Delhi: Narosa Publication House.

22. Komanduri, R., N. Chandrasekaran and L.M. Raff. 1998. Effect of tool geometry in nanometric cutting: A molecular dynamics simulation approach, *Wear* 219: 84–97.

23. Blackley, W.S. and R.O. Scattergood. 1991. Ductile regime machining model for diamond turning of brittle materials. *Precision Engineering* 13(2): 95–103.

24. Yan, J., K. Syoji, T. Kuriyagawa and H. Suzuki. 2002. Ductile regime turning at large tool feed. *Journal of Materials Processing Technology* 121(2–3): 363–372.

25. Lawn, B.R. and A.G. Evans. 1977. A model for crack initiation in elastic/plastic indentation fields. *Journal of Materials Science* 12: 2195–2199.

26. Arif, M., Z. Xinquan, M. Rahman and S. Kumar. 2013. A predictive model of the critical undeformed chip thickness for ductile–brittle transition in nano-machining of brittle materials. *International Journal of Machine Tools and Manufacture* 64: 114–122.

27. Xiandong, L. 2000. Ultra-precision turning technology. SIMTech Technical Report PT/00/008/PM.

28. Baltrao, P.A., A.E. Gee, J. Corbett and R.W. Whatmore. 1999. Ductile mode machining of commercial PZT ceramics. *Annals of the CIRP* 48: 437–440.

29. Bulla, B., F. Klocke and O. Dambon. 2012. Analysis on ductile mode processing of binderless, nano crystalline tungsten carbide through ultra precision diamond turning. *Journal of Materials Processing Technology* 212: 1022–1029.

30. Baumgartner. 1980. A statics and dynamics of the freely jointed polymer chain with Lennard-Jones interaction. *The Journal of Chemical Physics* 72(2): 871–879.

31. Carr, J.W. and C. Feger. 1993. Ultra precision machining of polymers. *Precision Engineering* 15: 221–237.

32. Casey, M. and J. Wilks. 1973. The friction of diamond sliding on polished cube faces of diamond. *Journal of Physics D: Applied Physics* 6(15): 1772–1781.

33. Wilks, E.M. and J. Wilks. 1972. The resistance of diamond to abrasion. *Journal of Physics D: Applied Physics* 5(10): 1902–1919.

34. Cheung, C.F. and W.B. Lee. 2000. Study of factors affecting the surface quality in ultra precision diamond turning. *Materials and Manufacturing Process* 15(4): 481–502.

35. Grzesik, W. 1996. A revised model for predicting surface roughness in turning. *Wear* 194(1): 143–148.
36. Vyas, A. and M.C. Shaw. 1999. Mechanics of saw-tooth chip formation in metal cutting. *Journal of Manufacturing Science and Engineering* 121(2): 163–172.
37. Weule, H., V. Hüntrup and H. Tritschler. 2001. Micro-cutting of steel to meet new requirements in miniaturization. *CIRP Annals-Manufacturing Technology* 50(1): 61–64.
38. Khan, G. S., S.V. Ramagopal, K.D. Chattopadhyay, P.K. Jain and V.M.L. Narasimham. 2003. Effects of tool feed rate in single point diamond turning of aluminium-6061 alloy. *Indian Journal of Engineering & Materials Sciences* 10(2): 123–130.
39. Zong, W.J., Y.H. Huang, Y.L. Zhang and T. Sun. 2014. Conservation law of surface roughness in single point diamond turning. *International Journal of Machine Tools and Manufacture* 84: 58–63.
40. Mishra, V., G.S. Khan, K.D. Chattopadhyay, K.N. and R.V. Sarepaka. 2014. Effects of tool overhang on selection of machining parameters and surface finish during diamond turning. *Measurement* 55: 353–361.
41. Juergens, R.C., R.H. Shepard III and J.P. Schaefer. 2003. Simulation of single-point diamond turning fabrication process errors. *Proceedings of SPIE, Novel Optical Systems Design and Optimization VI* 5174: 93–104.
42. Kong, M.C., W.B. Lee, C.F. Cheung and S. To. 2006. A study of material swelling and recovery in single point diamond turning of ductile materials. *Journal of Materials Processing Technology* 180(1–3): 210–215.
43. Zhu, Z. and S. To. 2015. Adaptive tool servo diamond turning for enhancing machining efficiency and surface quality of freeform optics. *Optics Express* 23(16): 20234–20248.
44. Neo, D.W.K., A.S. Kumar and M. Rahman. 2014. A novel surface analytical model for cutting linearization error in fast tool/slow slide servo diamond turning. *Precision Engineering* 38(4): 849–860.
45. Harvey, J.E., S. Schroder, N. Choi and A. Duparre. 2012. Total integrated scatter from surfaces roughness, correlation width and incident angle. *Optical Engineering* 51(1): 013402.
46. http://www.photonics.com/EDU/Handbook.aspx (accessed January 1, 2017).
47. http://www.precitech.com/products/nanoform250ultra/nanoform_250_ultra.html (accessed January 1, 2017).
48. http://www.diverseoptics.com/optics-materials (accessed January 1, 2017).
49. http://www.photonics.com/EDU/Handbook.aspx?AID=25504 (accessed January 1, 2017).
50. http://www.naluxnanooptical.com/clear-optical-plastics.html (accessed January 1, 2017).
51. Whitehouse, D.J. 2011. *Handbook of Surface – Nanometrology*, 2nd ed. Boca Raton, FL: CRC Press/Taylor & Francis.
52. Dagnall, H. 1997. *Exploring Surface Texture*. Leicester, England: Rank Taylor Hobson.
53. Amaral, M.M., M.P. Raelea, J.P. Caly, R.E. Samada, N.D. Vieira Jr. and A.Z. Freitas. 2009. Roughness measurement methodology according to DIN 4768 using Optical Coherence Tomography (OCT). *Proceedings of SPIE, Modeling Aspects in Optical Metrology II* 7390: 73900Z1–73900Z8.
54. Vorburger, T.V. and J. Raja. 1990. NIST Surface finish metrology tutorial. https://www.nist.gov/sites/default/files/documents/calibrations/89-4088.pdf (accessed January 1, 2017).

55. Novak, M. 2015. Non-Contact Surface Texture for Industrial Applications. https://www.bruker.com/fileadmin/user_upload/8-PDF-Docs/SurfaceAnalysis/3D-OpticalMicroscopy/Webinars/Non_Contact_Surface_Texture_-_Industrial_Applications.pdf (accessed January 1, 2017).

56. ASME B46.1-2009. 2010. Surface Texture (Surface Roughness, Waviness, and Lay). http://files.asme.org/Catalog/Codes/PrintBook/28833.pdf.

57. Optical Metrology Proceedings – Zygo Guide for surface texture parameters OMP-0514C. 2013. https://www.zygo.com/library/papers/SurfText.pdf (accessed January 1, 2017).

58. Mike Mills. 2011. Taylor–Hobson Tutorial – Cut-offs and the measurement of surface roughness. http://www.taylorhobsonserviceusa.com/uploads/2/5/7/5/25756172/tutorial_-_cut-offs_and_the_measurement_of_surface_roughness.pdf (accessed January 1, 2017).

59. Cohen, D. 2014. Michigan Metrology – Surface texture parameters glossary. http://www.michmet.com/Texture_parameters.htm (accessed January 1, 2017).

60. Whitehouse, D.J. 1982. The parameter rash – Is there a cure? *Wear* 83: 75–78.

61. Jenoptik Guide – Surface Roughness Parameters. 2013. https://www.jenoptik.com/cms/jenoptik.nsf/res/Surface%20roughness%20parameters_EN.pdf/$file/Surface%20roughness%20parameters_EN.pdf (accessed January 1, 2017).

62. LISA. 2002. Precision Devices Surface Metrology Guide – Surface Roughness. http://www.predev.com/smg/pdf/SurfaceRoughness.pdf (accessed January 1, 2017).

63. Khan, G.S., R.V. Sarepaka, K.D. Chattopadhyay, P.K. Jain and R.P. Bajpai. 2004. Characterization of nano scale roughness in single point diamond turned optical surfaces using power spectral density analysis, *Indian Journal of Engineering and Materials Science* 11: 25–30.

64. Gerchman, M.C. 1989. Optical tolerancing for diamond turning ogive error. *Proceedings of SPIE Reflective Optics II* 1113: 224–229.

65. Bittner, R. 2007. Tolerancing of SPDT diffractive optical elements and optical surfaces. *Journal of the European Optical Society – Rapid Publications 2*, 07028: 1–8.

66. Gerchman, M.C. 1986. Specifications and manufacturing considerations of diamond machined optical components. *Proceedings of SPIE* 607: 36–45.

67. Lamonds, D.L. 2008. Surface finish – Form fidelity in diamond turning. MS Thesis, North Carolina State University.

68. Yuan, W., W.B. Lee, C.Y. Chan and L.H. Li. 2016. Force and spatial profile analysis of surface generation of single point diamond turning. Proceedings of the 16th International Conference of the European Society for Precision Engineering and Nanotechnology, EUSPEN 2016.

69. Yuan, W., W.B. Lee, C.Y. Chan and L.H. Li. 2016. Development of a novel tool holder with six degree of freedom and the related tool path generation for ultra-precision machining, Proceedings of the 16th International Conference of the European Society for Precision Engineering and Nanotechnology, EUSPEN 2016.

70. Chan, C.Y., L.H. Li, W.B. Lee and H.C. Wong. 2016. Monitoring life of diamond tool in ultra-precision machining. *The International Journal of Advanced Manufacturing Technology* 82(5): 1141–1152.

71. Beyca, O.F., P.K. Rao, Z. Kong, S.T.S Bukkapatnam and R. Komanduri. 2016. Heterogeneous sensor data fusion approach for real-time monitoring in ultra-precision machining (UPM) process using non-parametric Bayesian clustering and evidence theory. *IEEE Transactions on Automation Science and Engineering* 13(2): 1033–1044.

72. Yao, H., Z. Li, X. Zhao, T. Sun, G. Dobrovolskyi and G. Li. 2016. Modeling of kinematics errors and alignment method of a swing arm ultra-precision diamond turning machine. *The International Journal of Advanced Manufacturing Technology* 87: 165–176.

73. Uhlmann, E., D. Oberschmidt, J. Polte, M. Polte and S. Guhde. 2015. New machine tool concept for two-side ultra-precision machining. Proceedings of the 15th International Conference of the European Society for Precision Engineering and Nanotechnology, EUSPEN: 353–354.

74. Huang, P., W.B. Lee and C.Y. Chan. 2015. Investigation of the effects of spindle unbalance induced error motion on machining accuracy in ultra-precision diamond turning. *International Journal of Machine Tools and Manufacture* 94: 48–56.

75. Tauhiduzzaman, M., A. Yip and S.C. Veldhuis. 2015. Form error in diamond turning. *Precision Engineering* 42: 22–36.

76. Brinksmeier, E., O. Riemer and L. Schönemann. 2015. High performance cutting for ultra-precision machining. *International Journal of Nanomanufacturing* 11(5–6): 245–260.

77. Dubrovinskaiaa, N., L. Dubrovinsky, W. Crichton, F. Langenhorst and A. Richter. 2005. Aggregated diamond nanorods, the densest and least compressible form of carbon. *Applied Physics Letters* 87: 083106.

78. A.L.M.T Corporation website: http://www.allied-material.co.jp/english/products /diamond/cutting/blupc/ (accessed January, 18, 2017).

79. Mir, A., X. Luo and J. Sun. 2016. The investigation of influence of tool wear on ductile to brittle transition in single point diamond turning of silicon. *Wear* 364–365: 233–243.

80. Schönemanna, L., O. Riemer and E. Brinksmeier. 2016. Control of a thermal actuator for UP-milling with multiple cutting edges. *Procedia CIRP* 46: 424–427.

81. Chan, C.Y., L.H. Li and W.B. Lee. 2015. Novel selection system of ultra-precision machining tool for optical lens. Proceedings of the 15th International Conference of the European Society for Precision Engineering and Nanotechnology, EUSPEN 2015: 315–316.

82. Doetz, M., O. Dambon, F. Klocke and O. Fähnle. 2015. Influence of coolant on ductile mode processing of binderless nanocrystalline tungsten carbide through ultraprecision diamond turning. *Proceedings SPIE, Optical Manufacturing and Testing XI* 9575: 95750R.

83. Doetz, M., O. Dambon, F. Klocke and O. Fahnle. 2016. Chemical influence of different pH-values on ductile mode processing through ultra-precision diamond turning. *Proceedings SPIE, Third European Seminar on Precision Optics Manufacturing* 10009: 1000905.

84. Chan C.Y., L.H. Li, W.B. Lee and H.C. Wong. 2016. Use of nano-droplet-enriched cutting fluid (NDCF) in ultra-precision machining. *The International Journal of Advanced Manufacturing Technology* 84: 2047–2054.

85. Chan, C.Y., W.B. Lee and H. Wang. 2013. Enhancement of surface finish using water miscible nano-cutting fluid in ultra-precision turning. *International Journal of Machine Tools & Manufacture* 73: 62–70.

86. Smith, P.J., B. Chu, E. Singh, P. Chow, J. Samuel and N. Koratkar. 2015. Graphene oxide colloidal suspensions mitigate carbon diffusion during diamond turning of steel. *Journal of Manufacturing Processes* 17: 41–47.

87. Neo, D.W.K. 2015. Ultra-precision machining of hybrid freeform surface using multiple-axis diamond turning. PhD Thesis, National University of Singapore.

88. Tian, F., Z. Yin and S. Li. 2016. A novel long range fast tool servo for diamond turning. *The International Journal of Advanced Manufacturing Technology* 86: 1227–1234.

89. Baier, K. 2016. New diamond turning strategy with 2-axis fast tool for dense dimple pattern on embossing rollers. Proceedings of the 16th International Conference of the European Society for Precision Engineering and Nanotechnology, EUSPEN 2016.

90. Zhu, Z., X. Zhou, D. Luo and Q. Liu. 2013. Development of pseudo-random diamond turning method for fabricating freeform optics with scattering homogenization. *Optics Express* 21(23): 28469–28482.

91. Yip, W.S., S. To and Y. Deng. 2015. Preliminary experimental study on ultrasonic assisted diamond turning Ti_6Al_4V alloy. Proceedings of the 15th International Conference of the European Society for Precision Engineering and Nanotechnology, EUSPEN 2015.

92. Lin, J., M. Lu and X. Zhou. 2016. Development of a non-resonant 3D elliptical vibration cutting apparatus for diamond turning. *Express Technologies* 40: 173–183.

93. Zhou, X., C. Zuo, Q. Liu and J. Lin. 2016. Surface generation of freeform surfaces in diamond turning by applying double-frequency elliptical vibration cutting. *International Journal of Machine Tools and Manufacture* 104: 45–57.

94. Zhang, S.J., S. To, G.Q. Zhang and Z.W. Zhu. 2015. A review of machine-tool vibration and its influence upon surface generation in ultra-precision machining. *International Journal of Machine Tools and Manufacture* 91: 34–42.

95. Neo, D.W.K., A.S. Kumar and M. Rahman. 2016. CAx-technologies for hybrid fast tool/slow slide servo diamond turning of freeform surface. *Proceedings of the Institution of Mechanical Engineers, Part B: Journal of Engineering Manufacture* 230(8): 1465–1479.

96. To, S., Z. Zhu and H. Wang. 2016. Virtual spindle based tool servo diamond turning of discontinuously structured microoptics arrays. *CIRP Annals – Manufacturing Technology* 65(1): 475–478.

97. Gong, H., Y. Wang, L. Song and F.Z. Fang. 2015. Spiral tool path generation for diamond turning optical freeform surfaces of quasi-revolution. *Computer Aided Design* 59: 15–22 (accessed January 1, 2017).

98. Huang, R., X. Zhang, M. Rahman, A.S. Kumar and K. Liu. 2015. Ultra-precision machining of radial Fresnel lens on roller moulds. *CIRP Annals – Manufacturing Technology* 64(1): 121–124.

99. Li, C.J., Y. Li, X. Gao and C.V. Duong. 2015. Ultra-precision machining of Fresnel lens mould by single-point diamond turning based on axis B rotation. *International Journal of Advanced Manufacturing Technology* 77(5): 907–913.

100. Neo, W.K., M.D. Nadhan, A.S. Kumar and M. Rahman. 2015. A novel method for profile error analysis of freeform surfaces in FTS/STS diamond turning. *Key Engineering Materials* 625: 101–107.

101. Otieno, T., K. Abou-El-Hossein, W.Y. Hsu, Y.C. Cheng and Z. Mkoko. 2015. Surface roughness when diamond turning RSA 905 optical aluminium. *Proceedings of SPIE, Optical Manufacturing and Testing XI* 9575: 957509.

102. Guan, C., H. Hu, G. Tie and X. Luo. 2016. A new process chain for ultra-precision machining potassium dihydrogen phosphate (KDP) crystal parts. Proceedings of the 16th International Conference of the European Society for Precision Engineering and Nanotechnology, EUSPEN 2016.

103. Abou-El-Hossein, K., O. Olufayo and Z. Mkoko. 2013. Performance of diamond inserts in ultra-high precision turning of Cu_3Cr_3Zr alloy. *Wear* 302: 1098–1104.
104. Wu, H.B. and S. To. 2016. Effects of electropulsing treatment on material properties and ultra-precision machining of titanium alloy. *The International Journal of Advanced Manufacturing Technology* 82: 2029–2036.
105. Mohammadi, H., D. Ravindra, S.K. Kode and J.A Patten. 2015. Experimental work on micro laser-assisted diamond turning of silicon (111). *Journal of Manufacturing Processes* 19: 125–128.
106. Han, J.D., W.B. Lee and C.Y. Chan. 2016. Establishment of a laser assisted ultra-precision machining system. Proceedings of the 16th International Conference of the European Society for Precision Engineering and Nanotechnology, EUSPEN 2016.

Index

Page numbers with f and t refer to figures and tables, respectively.

A

Abrasion, 70
Abrasive diamond powders, 51
Abrasive particle, for material removal, 36, 37f
Abrasive wear, 49, 51, 53
Accuracy of DTM; *see also* Diamond turn machines (DTM)
   balanced loop stiffness, 15–16, 15f
   positional accuracy, 13–15, 14f
   repeatability of moving elements, 13–15
   thermal effects, 16
   vibration effects, 16–17, 17f
Adhesion, 70, 71f
Aggregated diamond nanorod (ADNR), 135, 135f
Air-hammering effect, 135
Amorphous polymers, 40
Amplitude error, average, 117
Amplitude parameters, 115–119
Application programming interface (API), 140
Archimedes spiral, 86, 141
Aspherical diffractive optics, 98t
Aspherical lens, 98t
Asymmetric macro shapes, tool paths for slow tool servo (STA, 84–87
   synchonisation of spindle rotation, 83–84, 84f
Atomic force microscopy (AFM), 59

B

Bayesian Dirichlet process method, 133
Biomedical devices, 95
Brazing filler metals, 56
Brittle materials; *see also* Material removal mechanism
   about, 30
   ductile regime machining of, 39–40, 39t
   machining mechanism, 33–34, 34f

C

CAD software, 140
Carbon spots, 50
Characterisation, defined, 105
Chip formation, 34
Clamping method, 71, 72f
Coherence correlation interferometry (CCI), 57, 57f
Computer numerical control (CNC) motion paths, 47
Constant angle sampling strategy (CASS), 86, 87
Constant-arc-length sampling strategy (CLSS), 86, 87
Contact profilometer, 111, 111f, 112, 125
Controlled-tool motion, vibration-based, 138–140
Coolant in DTM, 70, 71f, 127, 137, 138
Crack formation, 34f
Crater wear, 57
Crystals, diamond turn machined, 102
Cutting edge surface, 48
Cutting mechanisms for engineering materials, 30–35, 33f, 33t; *see also* Material removal mechanism
Cutting tools
   and development, 135–137, 135f
   manufacturers and diamonds, 50

D

Damping, 4, 16, 18, 19, 20
Degree of freedom, 14, 14f
Depth of cut (DOC), 69, 69f
Deterministic finishing processes, 28, 28f

Deterministic surface generation, 90–92
Diamond
    structure of, 50, 51f
    tools, 53, 53f
    tool wear development, 58f
    turning technology, importance of,
        1–2
    turn machineable optical elements, 98t
Diamond turn machines (DTM)
    about, 11
    characteristics/capabilities of, 17–18,
        18f
    classification of, 11–12, 12f
    components of, 19, 19f, 20
    environmental conditions for, 21–22
    requirements of
        balanced loop stiffness, 15–16, 15f
        overview, 12–13, 12f
        positional accuracy/repeatability
            of moving elements, 13–15, 14f
        thermal effects, 16
        vibration effects, 16–17, 17f
    solved problems (sample), 22–25
    specification of (sample), 22, 23t
    technologies in, 20, 21t
Diamond turn machining (DTM)
    advances in
        controlled-tool motion,
            vibration-based, 138–140, 140f
        coolant in DTM, 137–138, 138f
        cutting tools and development,
            135–137, 135f, 136f
        machine tools and development,
            133–135, 134f
        materials and materials treatment,
            142–144
        process monitoring, 131–133, 132f
        tool holding for DTM, 144
        tool-path planning techniques,
            140–142, 141f, 142f
    applications
        application areas of, 101–102, 101t
        categories, 95–96
        crystals, 103
        diamond turn machined ultra-
            precision components, 101–102,
            102f
        infrared and near infrared optics,
            100, 101f

        materials machinable by DTM,
            102–103
        metal optics, 100, 100f
        metals, 102
        mold inserts for polymer optics,
            99–100
        in optical domain, 96–99, 98t
        polymer optics products, 99, 99f
        polymers, 102
    defined, 5
    development of, 7–8
    dynamics of, 8
    optimisation of DTM parameters,
        74, 75t
    process and parameters
        about, 63–65, 64f
        clamping method/footprint error,
            71–72, 72f
        coolant, 70–71, 71f
        depth of cut (DOC), 69, 69f
        feed rate, 67–68, 67t, 69f
        spindle speed, 65–66, 66f, 66t
        tool shank overhang, 69, 70f
    in process chain, 8–10, 9f
    surface quality, 5–7
    thermal issues in, 73–74
    uniqueness, 5
    vibration related issues, 72–73, 73f
Diffraction gratings, 98t
Drive systems, 19f, 19t
DTM (Diamond turn machines), *see*
        Diamond turn machines
        (DTM)
DTM (Diamond turn machining), *see*
        Diamond turn machining
        (DTM)
Ductile material
    about, 30
    factors affecting removal of, 32
    machining mechanism, 31
Ductile regime machining of brittle
        materials, 39–40, 39t

E

Electroless nickel, 99, 102
Electromagnetic (EM) waves
    fabricating components, 3
    for surface smoothness, 1, 2f

Electro-pulse treatment (EPT), 143
Elliptical vibration cutting, 139, 140f
Engineering materials, classification, 30
Error compensation techniques, 128–129, 128f
Errors in surface quality, 122
Evaluation length, 114
External vibration, 16

F

Fast tool servo (FTS) system, 54, 65, 84, 88–89, 88f 138
Feed motion, 4, 6, 10, 80, 81, 83
Feed rate, 67–68, 67t, 69f
Figure error, 109, 129
Filler metals, 56
Filter in surface characterization, 113
Finish error, 109, 110
Flexible clamping, 71
Footprint error, 64, 71, 74
Form error, 108, 109, 129
Freeform optics, 98, 99
Fresnel lens, 98, 141
Friction coefficient, 51

G

Gaussian thermal profile, 127
Germanium optics, 100
Glass transition temperature, 40
Ground vibration, transmission of, 22

H

Heat energy, 73
Heat generation, damages on surface integrity, 73
Heat transfer, 70
Height parameter, 116
High spatial frequency (HSF), 91
Holding errors during DTM, 125, 125f
Humidity control, 21

I

Infrared (IR) waves, and DTM, 9
Infrared optics, 100, 101f

J

Joint stiffness, 15

K

Kinematic error model, 133

L

Lapping process, *see* Random finishing processes
Lathe machines, 11
Lay pattern, 90, 90f
Lenard–Jones potential, 41
Lenslet array, 98t
Light scattering tests, 139
Loop stiffness, 15, 16
Low spatial frequency (LSF), 91

M

Machine tools and development, 133–135, 134f
Machining processes, 4, 5
Material-induced vibration, 16, 17
Material removal mechanism
  by abrasive particle, 37f
  cutting mechanisms for engineering materials, 30–35, 31f, 32t
  deterministic/random machining process, comparison, 28–30
  ductile regime machining of brittle materials, 39–40, 39t
  micro-/nano-regime cutting mechanisms, 35–39
  overview, 27–28
  polymers, machining of, 40–41, 40f
Material removal rate (MRR), 29
Materials machinable by DTM, 102–103
Metals
  diamond turn machined, 102
  molds, 99, 99f
  optics, 100, 100f
Metrology
  defined, 105
  errors, 124–127, 125f, 126f, 127f
  by stylus-based profilometres, 121
  thermal effects and, 127

Micro-optics arrays (MOA), 141
Micro-regime cutting mechanisms,
        35–39, 35f, 36f, 37f, 39f; *see also*
        Material removal mechanism
Mid-spatial frequency (MSF), 91–92, 91f
Milling, 53, 136, 136f
    machines, 11
    ultra-precision, 55
Milling-type intermittent motion, 6
Mold inserts for polymer optics, 99, 99f
Molecular dynamics simulation (MDS)
        technique, 36, 38f
Monomers, 41
Multigrain machining, material
        removal, 34, 35f

N

Nano-droplet-enriched cutting fluids
        (NDCF), 137
Nano-regime cutting mechanisms,
        35–39, 35f, 36f, 37f, 39f; *see also*
        Material removal mechanism
Natural crystal growth, 50
Natural diamonds, 49, 50, 56
Near infrared optics, 100, 101f
Newton's second law of motion, 36

O

Ogive error, 123–124, 12f4, 124t
Optical components, process chain for, 9
Optical scattering and surface
        roughness, 96
Optics
    classifications of, 96
    optical elements, 97, 98
Optimisation of DTM parameters, 74
Over-cut ogive error, 123

P

Parameters in surface characterization,
        113
Peak-to-valley profile amplitude error,
        116
Peak-to-valley value, 120, 121
Plastic deformation, 32, 39, 137
Polishing process, 9, 53

Polycrystalline cubic boron nitride
        (PCBN), 48
Poly crystalline diamond (PCD), 48, 49f
Polymer optics, 96, 99, 99f
Polymers
    diamond turn machined, 102
    machining of, 40–41, 40f; *see also*
        Material removal mechanism
Potassium dihydrogen phosphate
        crystals, 143
Power spectral density (PSD), 72, 73, 91,
        119
Precision component production cycle,
        106f
Precision conic surfaces, 108
Precision machines (PM), 13
Precision optics, 101
Precision surfaces, 105–108
Preston's equation, 29
Primary profile in surface
        characterization, 113
Problems
    solved (sample)
        diamond turn machines, 22–25
        DTM parameters, 75–77, 76f, 77f
        material removal mechanism, 42–44
    unsolved
        diamond turn machines (DTM), 25
        material removal mechanism, 45
        tooling for DTM, 61
Profile error analysis (PEA), 142
Profile error compensation, 128, 128f
Profile in surface characterization, 113
Profilometres, 113, 121

R

Radius of curvature (RoC), 108
Random finishing processes, 28, 29
Relaxation time, 40, 41
Repeatability of moving elements, 13–15
Resonant systems, 139
Rigid clamping, 71, 72
Root-mean-squared (rms) value, 120, 121
Rotationally asymmetric shape, DTM,
        79, 80f
Rotationally symmetric shape, DTM,
        79, 80f
Rough cut motion paths, 83

Roughness error, average, 118
Roughness in surface characterization, 113

S

Sampling length, 115
Scanning electron microscopy, 57
Sensor fusion approach, 133
Silicon diffractive optics, 100f
Silicon optics, 100f
Single axis FTS systems, 139
Single crystal (SC)
    about, 49
    anisotropic properties, 50
    material, 6
Single crystal diamond (SCD), 39, 135
    tools, 49–53, 51f, 55f; *see also* Tooling for DTM
Single point diamond turning (SPDT), 8
Single point machining process, 36
Slow tool servo (STS), 54, 84–87, 85f, 86f, 138
Smooth surface, engineering applications and, 1–4, 2f, 3f, 3t
Space technology and diamond turning technology, 3
Spatial parameters, 115; *see also* Surfaces-metrology-characterisation
Specification of diamond turn machine (sample), 22, 23t
Specific cutting energy, 31, 33, 36
Spherical lens, 98
Spindle rotation, synchronisation of, 83–84, 84f
Spindle speed, 65–66, 66f, 66t
Spindle vibration, 16
Spiral motion path concept, 83
Stylus-based profilers, 121
Stylus instrument, 114
Sub-grain material removal, 35
Surface
    defined, 113
    errors, quantification of, 109, 110
    grinding, 90
Surface finish, values in DTM, 67, 67t
Surface generation, tool path strategies in asymmetric macro shapes
    slow tool servo (STS), 84–87, 85f, 86f

    synchronisation of spindle rotation, 83–84, 84f
    deterministic surface generation, 90–92, 90f, 91f, 92f
    micro-features, producing
        about, 87, 88f
        fast tool servo (FTS), 88–89, 88f
    overview, 79, 80f
    symmetric macro shapes, 80–83, 80f, 81f
    tool-normal motion path, 89–90, 89f
Surface quality
    figure error, 109
    finish error, 109
    form error, 108, 109
    sources of errors in, 122
Surface roughness
    along cutting velocity path, 132
    amplitude parameters, 119
    feature, 113
    measurement, 121
    optical surface, 98
    sub-nanometric roughness, 53
    and surface profiler, 91
Surfaces-metrology-characterisation
    amplitude parameters, 115–119
    characterisation, defined, 105
    error compensation techniques, 128, 129
    errors in surface quality, sources of, 122
    metrology
        defined, 105
        errors, 124–127
        ogive error, 123, 124
        by stylus-based profilometres, 121
        thermal effects and, 127
    power spectral density (PSD), 119
    precision surfaces, 105–108
    spatial parameters, 115
    surface errors, quantification of, 109, 110
    surface quality
        figure error, 109
        finish error, 109
        form error, 108, 109
    surface texture
        about, 110–112
        parameters, 112–115
    tolerance, 120

Surface texture, 110–112, 110f, 111f,
    parameters, 112–115, 114f, 115f
Surface topography generation
    algorithm, 139
Swelling of material, 73, 127
Symmetric shapes, tool motion path for,
    80f
Synchronisation of spindle rotation, 83,
    84
Synthetic diamonds, 50, 56

T

Taper scratch tests, 137
Telescopes, 3
Temperature control, 21
Thermal drift, on DTM, 16
Thermal flux, 127
Threshold chip thickness, brittle
    materials, 39
Titanium alloys, 143
Tolerance, defined, 120
Tool holding for DTM, 144
Tooling for DTM
    diamond tool fabrication, 55–57, 55f
    overview, 47
    single crystal diamond (SCD) tools,
        49–53, 51f, 52f
    tool geometry, 53–55, 53f, 54f, 55f
    tool materials, 47–49, 48f, 49f
    tool setting in DTM, 60, 60f
    tool wear, 57–58, 58f, 59f, 60f
Tool-normal motion path, 89–90, 89f;
    see also Surface generation,
        tool path strategies in
Tool-path compensatory approach, 128
Tool-path planning techniques, 140–142,
    141f, 142f
Tool setting error, 59, 60
Tool shank overhang, 69, 70f
Tool tip vibration, 15
Tool wear, 68, 73
Total integrated scattering (TIS), 98
Touch-probe profilers, 121
Traced profile, 114
Traversing length, 114
Turning process, *see* Deterministic
    finishing processes
Type A machines, 11, 12f

Type B machines, 11, 12f
Type C machines, 11, 12f
Type D machines, 11, 12f

U

Ultra-precision components, diamond
    turn machined, 101
Ultra-precision machining (UPM),
    5, 8; *see also* Diamond turn
        machines (DTM)
Under-cut ogive error, 123
Unfiltered primary profile, 114, 114f
Unit removal (UR) of material, 29, 67

V

Vacuum chuck clamping, 71, 72
Van der waal's forces, 41
Vaviness profile, 113
Verlet algorithm, 38
Vibration
    elliptical vibration cutting and, 139
    at interface of DTM tool, 16, 17
    isolators, 22
    material-induced, 16–17, 17f
    sources of, 72
Viscoelasticity, 41

W

Waviness
    about, 125, 128
    determination, 57, 58f, 59f
    in surface characterization, 113
W-axis, short-stroke, 84
Wear prediction models, 65

X

X-ray beam deflections, 1
X-ray mirror, 1, 3f
X-ray photoelectron spectroscopy (XPS),
    138

Z

Z-axis slide system, 84, 134
Zero error, 129